THE
Mythical
Zoo

THE MYTHICAL ZOO

Animals in Myth, Legend, and Literature

BORIA SAX

OVERLOOK DUCKWORTH

NEW YORK • LONDON

This edition first published in hardcover in the United States and the United Kingdom in 2013 by
Overlook Duckworth, Peter Mayer Publishers, Inc.

NEW YORK:
141 Wooster Street
New York, NY 10012
www.overlookpress.com
For bulk and special sales, please contact sales@overlookny.com,
or write us at the address above

LONDON:
30 Calvin Street
London E1 6NW
info@duckworth-publishers.co.uk
www.ducknet.co.uk
For bulk and special sales, please contact sales@duckworth-publishers.co.uk,
or write us at the address above

Cataloging-in-Publication Data is available from the Library of Congress

Sax, Boria.
Mythical zoo : animals in myth, legend, and literature / Boria Sax.
pages cm
Includes bibliographical references and index of animals.
1. Animals and civilization. 2. Animals—Symbolic aspects. 3. Animals, Mythical.
4. Animals—Folklore 5. Animals in literature. I. Title.
QL85.S27 2013   590—dc23   2013033492

Book design and typeformatting by Bernard Schleifer
Manufactured in the United States of America
ISBN US: 978-1-4683-0720-7
ISBN UK: 978-0-7156-4732-5

To Humankind—

So endlessly vulnerable, foolish, guilty, fragile . . .
Yet as wondrous as the other creatures in this book.

Now as at all times I can see in the mind's eye,
In their stiff, painted clothes, the pale unsatisfied ones
Appear and disappear in the blue depth of the sky
With their ancient faces like rain-beaten stones,
And all their helms of silver hovering side by side,
And all their eyes still fixed, hoping to find once more,
Being by Calvary's turbulence unsatisfied,
The uncontrollable mystery on the bestial floor.

—W. B. YEATS, "The Magi"

CONTENTS

PREFACE

BOTH ANIMALS AND THEIR SYMBOLISM ARE OVERWHELMING IN THEIR variety, and this book could theoretically become as endless as the subject itself. In a similar way, the sources of this book are so many and varied that it would be very cumbersome, and perhaps impossible, to list them all. In a few retellings of stories, I have taken the liberty of inventing bits of dialogue to make the story more vivid. In doing this, I have never changed the plot, and this mild license is certainly in the tradition of the legendary Aesop, whose stories exist in several variants but not in any definitive edition.

In a project such as this one, finding material is not a problem, but deciding what to leave out can be a very big one indeed. My policy has been to emphasize depth rather than breadth. I wished to convey the underlying ideas behind the treatment of animals in myth, legend, and related aspects of human culture, rather than simply give bits of disconnected information.

For the help I've received in putting together this volume, there are many to acknowledge. I wish to thank my wife Linda Sax, who meticulously read over the manuscript, made corrections, and offered many splendid suggestions. In addition to a fine command of written English, she brought a special perspective as a living history teacher at Historical Hudson Valley. My agent, Dianne Littwin, was extremely helpful in making arrangements, and I also wish to thank her for her faith in the work. I want to thank Peter Mayer and the team at the Overlook Press for their interest, initiative, and helpfulness.

—BORIA SAX, *June 2013*

Moonmoth and grasshopper that flee our page
And still wing on, untarnished of the name
We pinion to your bodies to assuage
Our envy of your freedom—we must maim
Because we are usurpers, and chagrined—
And take the wing and scar it in the hand.
Names we have, even, to clap on the wind;
But we must die, as you, to understand.

HART CRANE, "A Name for All,"

INTRODUCTION:
ANIMALS AS TRADITION

OR CENTURIES, JUST ABOUT EVERYBODY HAD ASSUMED THAT BATS WERE mice with wings. When Linnaeus refined the Aristotelian classification of animals in the eighteenth century, he challenged this "common sense," in the name of an eternal order decreed by God. After carefully examining their anatomy, Charles Linnaeus proclaimed that bats were actually primates, like monkeys and human beings. Decades later, he reconsidered and placed bats in their own category, the *chiroptera*, where they have stayed ever since. A bit less than a century later, Darwin's Theory of Evolution provided taxonomists with a new paradigm—that of biological inheritance.

But to define an animal strictly in terms of evolution is too narrow, technical, reductionist, or restrictive for many purposes. Scientists generally regard animals as belonging to different "species" when they do not habitually mate with one another. Although dogs, wolves, jackals, and coyotes are capable of mating together, they generally do not do so in the wild, so each of these is considered a distinct type. This biological definition loses much of its meaning under conditions of domestication, whether on a farm or in a zoo, where animals do not necessarily choose the partners with which they breed. A horse and an ass can mate, and they are often induced to do so in order to produce a mule, which retains useful qualities of each.

The definition becomes almost entirely meaningless when dealing with animals produced by genetic engineering, for which one cannot really speak of "species" at all. Scientists have produced a cross between a sheep and a goat, known as a "geep." They have placed human genetic material in pigs, in order to produce organs that will not be rejected when transplanted into human beings. They also have produced laboratory rats with transparent skin, so that their organs can easily be observed during experiments. Some of these creatures are like the monsters of folklore, and it may be that in the future we will see crosses between human beings and apes or dogs.

With gene splicing, it is now possible to cross not only the divisions of species, but even those between plants and animals. Scientists have inserted genes from flounders into the genetic code of tomatoes in order to increase their resistance to frost, and they have introduced genes from chickens into tomatoes to make the plants more resistant to disease. By inserting genes from a jellyfish into tobacco, they have produced plants that glow in the dark. Genetic theory views all living things, from ferns to human beings, less as either individuals or representatives of species than as repositories of hereditary information, to be endlessly recycled in new combinations.

One alternative method of classification is to lump creatures together by habitat. The old mariners who considered the whale a fish (rather than a mammal) really had a point. Just as we, for the most part, classify most of the people who live in France as "French," regardless of race, height, complexion, age, temperament, and so on, we can classify animals that live in the woods as "creatures of the forest." Shared environment arguably creates a more intimate sort of kinship than genetic inheritance, since it involves interaction and common experience. Human beings usually feel greater affinity with dogs and cats, which often share our homes, than we do with the great apes, our closest biological relatives. People have been learning from pets for millennia, and these animals have been learning from us. That is why dogs are vastly better able to interpret human body language than apes. Dogs understand the pointing of a human finger without being taught, while apes are almost always unable to comprehend this gesture.

But suppose we work to preserve either a species in the wild or a breed in domesticity. What exactly are we trying to perpetuate? A collection of physical characteristics? A bit of genetic code? Part of a habitat? If we define

Animals entering a steam ark.

(Illustration by J. J. Grandville from Un Autre Monde, 1847)

each sort of animal as a tradition, our definition includes all of these and more. It also includes stories from myth, legend, and literature. Such tales, with the love and fear they may engender, are part of an intimate relationship with human beings that has been built up for many centuries.

To regard each sort of animal as a tradition also encourages respect. Why should we care about species extinction? Appeals to transcendent reasons do not easily satisfy people in our secular society. Appeals to pragmatic reasons, such as preserving the ecosystem, are easily subject to challenge. Tradition robustly links animals not only to their natural environments but also to cultural values and practices over millennia—indeed, to our identity as human beings. It reveals how environmental interdependence is spiritual as well as physical. Without other creatures, "humanity," as we know it, would perish, even if our genetic inheritance continued to be passed on.

My Merriam-Webster dictionary defines "tradition" as "an inherited pattern of thought and action." It comes from "trade," which originally meant "track." To study a tradition is to track a creature, as though one were a naturalist, back through time. In the chapters that follow, I must at times ask the reader to make a special effort of imagination, and to forget for a while, as much as is reasonably possible, what he thinks he "knows" about the animals in question. Try to imagine each as it might have been experienced in unfamiliar cultures and environments. Finally, consider every animal as a sort of primal experience that reflects, creates, challenges, and, to a degree, transcends the limits of culture. These are animals of the spirit, living in the geography of the mind.

Ponder what it is like to be a bat and navigate by echolocation, i.e., by sonar. I imagine that must resemble entering a sort of spirit world, where things have definite locations but lack solidity. What is it like to be a dog, with a sense of smell 500 times as strong as a human being's? Perhaps the scents must be rather like intense intuitions, precise and yet not quite tangible. What is it like to be a shark, and hunt prey by their electromagnetic fields? Maybe the experience is akin to living in a musical world, where everything is better expressed in notes than in words. All of these possibilities have their counterparts in human culture.

No person is ever completely "human," and no animal entirely lacks humanity. We discover qualities in animals that we wish to lay claim to, and others which we attempt to disavow. Human beings construct our identity—collective, tribal, and individual—largely by reference to animals. Other creatures not only make us "human," as Paul Shepherd has observed, but also divide us into groups, making us females, males, Japanese, Americans, Mexicans, Christians, Buddhists, artists, intellectuals, mechanics, warriors, saints, criminals, and so on. This process is attested by the ubiquity of animal imagery in most of the symbols and stories that define our heritage as human beings. We have merged with animals through magic, metaphor, or fantasy, growing their fangs and putting on their feathers, to become deities, sages, tricksters, devils, clowns, companions, lovers, and far more.

1
ALMOST HUMAN

THE DISTINCTION BETWEEN "HUMAN" AND "ANIMAL" IS BY NO MEANS universal to all cultures. Many languages, such as Classical Chinese, have no equivalent of our Western concept of "animal." The closest thing to the concept of "human" among the Chewong, who live in forests of Malaysia, is a continually shifting sense of affinity, which may at times include trees or insects, while excluding some men and women. And even when such a distinction is present, it is not necessarily conceived in ways that are familiar to us. The Karam people of New Guinea regard the cassowary—a large, flightless bird—as human. Many indigenous people, as well as quite a few European royal and noble families, trace their genealogies to animals in myths. Modern Western culture is probably unique in the central role it accords to the distinction between "civilization" and "nature," as well as to subsidiary distinctions such as the one between "people" and "animals," but, even in the Occident, that differentiation has never been stable or unambiguous. Westerners have, at times, partially excluded groups such as Black Africans from "human" status. Several other creatures have, for limited historical periods, been granted this status, at least in many contexts. In the ancient world, foremost of these was the bear, and, in the peasant culture of Medieval Europe, that became the pig. For a relatively brief period in the eighteenth and nineteenth centuries, many people thought the beaver was the animal closest to being human. With the acceptance of Darwin's theory of evolution, it became the ape.

Another animal that might have been included here is, of course, the dog, except for its relative lack of autonomy in human society. It will be covered in Chapter 9, "Man's Best Friends."

APE AND MONKEY

How like to us in form and shape is that ill-favored beast, the ape.
Attributed to Ennius by Cicero, *De Natura Deorum*

For most of history, people have distinguished between human beings and apes at least as much by habitat as by appearance or biology. Human beings lived in houses, usually in urban centers or the countryside, while apes lived in trees. Folklore also makes the forest home to fairies, elves, and other nature spirits, which are often given simian features such as long arms and profuse body hair.

In the early sixth century BCE, the Carthaginian navigator Hanno led an enormous expedition down the West Coast of Africa. According to his account of the voyage, they sighted a huge mountain called the "chariot of the gods." He and his crew confronted wild men and women covered with hair, which threw stones at them, and climbed adroitly up the slopes. Ever since then, people have speculated about whether the "wild men" were people dressed in animal skins, chimpanzees, baboons or, most likely, gorillas. Similar rumors of wild men have continued to circulate ever since, as mariners in distant lands reported seeing men with the heads of dogs, the feet of goats, their faces on their chests, and many creatures just as strange. Stories of hairy, wild men with clubs were told around the fire and sometimes acted out in Medieval pageants.

Among the most popular religious figures of China is Old Monkey, who was born when lightning struck a stone. Old Monkey broke into heaven, got drunk on celestial wine, erased his name from the book of the dead, and fought back the armies of heaven. Finally, he caused so much trouble that the gods and goddesses appealed to the Buddha himself for help. Buddha sought out Old Monkey, and their dialogue went something like this:

Illustration of various monkeys from a nineteenth-century book of natural history

"What do you wish?" the Buddha asked Old Monkey.

"To rule in heaven," Old Monkey replied.

"And why should this be granted to you?" asked the Buddha.

"Because," Old Monkey said, "I can leap across the sky."

"Why" laughed the Buddha, "I will bet that you cannot even leap out of my hand." He picked Old Monkey up, and said, "If you can do that, you may rule in heaven, but, if you can't, you must give up your claim."

Old Monkey made a tremendous jump and soon arrived at a pillar that holds up the sky. To show that he has been there, he urinated and wrote his name. Then, with another leap, Old Monkey returned to claim his prize.

"What have you done?" asked the Buddha.

"I have gone to the end of the universe," said Old Monkey.

"You have not even left my hand," laughed the Buddha, and he raised one finger. Recognizing the pillar of heaven, Old Monkey knew that the Buddha was right.

The Buddha imprisoned Old Monkey under a mountain for five hundred years. Finally, Old Monkey was rescued by Kwan-Yin, Bodhisattva of mercy. To achieve redemption, Monkey had to guard a Buddhist monk on a dangerous journey from China to India. Monkey gave the monk faithful, though not unwavering, service through fantastic adventures in which Monkey battled countless demons and sorcerers, to finally become a Buddha himself at the end. These adventures were chronicled in the epic novel *Journey to the West*, attributed to Ch'en-eng Wu, in the early sixteenth century. Old Monkey was often depicted alongside solemn portraits of Buddhist sages, but even as a Buddha he retained a mischievous streak.

The monkey-god Hanuman is similarly beloved in the Hindu pantheon, largely because he is capable of both childish mischief and noble sacrifice. When Hanuman was a child, he looked up, saw the sun and thought it must be a delicious fruit. He jumped to pick it and rose so high that Indira, god of the sky, became angry at the invasion of his domain. Indira hurled a thunderbolt at the intruder, striking Hanuman in the jaw. At this, the father of Hanuman, Vayu, god of winds, became furious, and started a storm that soon threatened to destroy the entire world. Brahma, the

Renaissance illustration showing an apelike creature allegedly captured in 1530 in Saxony

supreme god, placated Vayu by granting Hanuman invulnerability to weapons. Indira then added a promise that Hanuman could choose his own moment of death. Ever since, however, monkeys have swollen jaws. This story, from The *Ramayana*, an ancient Hindu epic, shows the amusement with which apes have generally been regarded throughout the world. Because Hanuman is a monkey, his divinity does not seem intimidating. In the Hindu *Panchantantra* and the early Buddhist *Jatakas*, the ape was one of the more sensible of animals, often a chief advisor to the lion king. People of the Far East regard the playfulness of monkeys and apes as divine serenity, not simple frivolity as in the West.

To find a major simian figure in Western religion, we must go all the way back to Thoth, the baboon-headed god of the ancient Egyptians. Thoth was the scribe to Osiris, ruler of the dead, and inventor of the arts and sciences. Perhaps in those archaic times reading and writing, still novel and full of mystery, appeared more simian than "human." Today, of course, we invoke language, especially writing, to proudly distinguish ourselves from all other creatures.

By contrast, the peoples of Mesopotamia and Greece often regarded apes as degenerate human beings. According to one Jewish legend, some of those who built the Tower of Babel were turned into apes. According to another, apes were the descendants of Enosh, the first son of Seth and grandson of Adam. On the other hand, some legends also told that Adam had a tail like that of a monkey. The debate over whether apes should be considered human probably goes back to the beginnings of civilization. In the religion of Zoroaster, the monkey or ape is the tenth and lowest variety of human beings created by Ahura Mazda.

In many mythologies, apes or monkeys were created as alternative human beings. In Philippine mythology, Bathala, the Creator of the world, was lonely and decided to make the first human being out of clay. When he was almost finished, the lump of clay slipped from his hand and trailed to the ground. That created a tail, and the figure became a monkey. Bathala created people on his second try.

According to the mythology of the Maya, the Creator once tried to fashion people of wood. They behaved so wickedly that all the animals and deities turned against them. The few people that remained retreated into the forest and became howler monkeys, and then the Creator made new human beings out of maize.

Wild men and women are intermediary beings related to both people and apes, at least in folklore. Among the earliest was Enkidu in *The Epic of Gilgamesh*, a heroic epic from Mesopotamia from the early second millennium BCE. Being created not by human parents but out of clay by the gods, he jostled with the beasts at the watering hole. With more than human strength, he overturned the traps of hunters. All who saw him were filled with fear and awe. When his subjects appealed to King Gilgamesh for help, he sent a prostitute to Enkidu, and she taught the wild man the ways of men. Enkidu drank wine instead of water, and he began to dress as a human being. But then the beasts rejected him, and human sorrow slowed his step.

Body hair, especially on men, is traditionally a sign of wildness. In the Bible, Esau, eldest son of Isaac and brother of Jacob, was covered with hair. He was also a bit of a wild man, one who liked open country and hunting but was ready to sell his birthright for a bowl of soup. When Isaac was old and blind, he prepared to bless Esau. Aided by his mother, Jacob covered himself up with the fleece of lambs. Jacob then went to his father, pretending to be Esau. Isaac demanded to touch his son. Fooled by the fur, Isaac allowed Jacob to steal his brother's benediction (Genesis 27). The story has often been interpreted as a triumph of civilization over savagery.

In the Islamic world as well, the resemblance of apes to human beings made them appear disconcerting, and they were often invoked to mock or parody human beings. The anonymous Medieval *Arabian Nights Entertainments* contained a story in which a cruel Jinni, finding his mistress in the company of a man, killed the woman and turned her companion into an ape. The man wandered in simian form until he came to the court of a king, who was amazed at his skill at calligraphy and chess. The King proudly ordered the ape to be dressed in fine silk and be fed on rare delicacies. A eunuch summoned the princess, so that she as well might see the wondrous animal. On entering the room, the princess immediately veiled her face, for she, as a Muslim, considered it improper that a strange man should see her features. She explained to her father that, without his knowledge, she had studied under a wise woman, was herself a great enchantress, and knew that the visitor was not an ape but a man. The King commanded his daughter to disenchant the ape, so that he might make that man his vizier. The Jinni appeared, his eyes burning like torches, and the princess began to recite some magic words. As the two traded spells, the Jinni became a lion, a scorpion, and then an eagle; the princess became a serpent, a vulture, and then a cock. They fought underneath the ground, in water and in fire, until at last the Jinni was burned to ashes. The princess as well received a mortal wound, but she was able to disenchant the ape before she died.

The Barbary apes (which are not really "apes," but macaques) were once popular pets of nobles and of wandering entertainers in Medieval Europe. They were probably originally brought from Africa by the Moors, Spaniards, or Portuguese, though nobody knows exactly when. Feral Barbary apes were found scattered near the Mediterranean coast into modern times, and a small population still hangs on at the Rock of Gibraltar. The sight of them vanishing into the trees may have contributed to many legends of fairies and wild men. According to a popular saying, Britain will fall if the apes ever disappear from Gibraltar.

The word "monkey" was probably first used to refer to macaques. While the etymology is uncertain, it may originally have been an affectionate diminutive meaning "little monk." Renaissance painters such as Albrecht Dürer often included monkeys in religious and courtly paintings, adding a playful touch to otherwise solemn occasions. In the Rococo art of the eighteen century, especially in France, pet macaques were often used to add a note of lightness and gaiety to scenes of luxurious rooms and gardens.

In Early Modern times, an expansion of maritime trade and exploration took Europeans to exotic corners of the world. Explorers began to discover both the great apes and people of cultures radically different from their own; sorting the former from the latter was not an easy matter. Scientists as well as sailors often conflated orangutans with gorillas and African tribesmen, all of whom were known mostly through fleeting glimpses and rumors. Tribes in West Africa regarded apes as human. Some believed that chimpanzees could speak but chose not to, so that they would not be forced to work. "Orangutan" was initially a Malay word for "wild man." When the Dutch anatomist Nicolaas Tulp dissected a body of an orangutan in 1641, he thought that it was the satyr of classical mythology. A colleague of his, Jacob de Bondt, believed these creatures were "born of the lust of Indonesian women who consort in disgusting lechery with apes."

Explorers in the sixteenth and seventeenth centuries brought tales back to Europe of apes that lived in huts, foraged in trees, and fought with cudgels. According to some accounts from the period, apes ravished human females or made war on human towns. The enormously popular *History of Animated Nature* published by Oliver Goldsmith during the late eighteenth century reported that, in Africa, apes sometimes steal men and women to keep as pets. Visitors to Victorian zoos complained that the apes tried to seduce human women. Sometimes apes were even made to put on clothes.

Literature as well blended folklore of wild men and women with recent accounts of primitives and other primates. In *Gulliver's Travels* by Jonathan Swift (1726), the hero was marooned on an island and adopted by highly civilized horses. In the woods on the fringes of their settlement were hairy men and women known as "yahoos." These primates constantly wallowed in their own filth. They had long claws and swung through trees. They roared, howled, and made hideous faces. The narrator was filled with revulsion at them, yet he could not help but acknowledge that these creatures were his own kind. The reaction clearly showed the feelings of Europeans at the gradual realization of their kinship with apes. This disgust of Gulliver with the yahoos anticipated the racism that took such terrible forms in the next few centuries. He reported, with fear yet no suggestion of disapproval, that the horses proposed a complete extermination of yahoos.

Apes would later figure prominently in racist propaganda. We can see the beginning of this in the story "The Origin of Apes," told by the late Medieval German folk poet and shoemaker Hans Sachs. Jesus, accompanied by Peter, was wandering through the countryside and stopped at a blacksmith's house. Along came an elderly cripple, and Peter asked Jesus to make the invalid young and strong. Jesus promptly consented, and he told the smith to heat his furnace. When the fire had started to blaze, Jesus placed the cripple inside it, and the man glowed with light. After saying a blessing, Jesus took the man out of the flames, and dipped him in water; everyone was amazed to see the cripple transformed into a strong young fellow. After Jesus left, the smith's elderly mother-in-law wanted to be rejuvenated as well. The smith, who had watched everything, agreed. After placing the old woman in the furnace, just as Jesus had done with the cripple, the smith realized that the magic was not working properly. He pulled his screaming mother-in-law out of the tub and dipped her in water. The cries summoned the smith's wife and daughter-in-law, both of whom were pregnant; they saw the wrinkled, distorted face of the howling old woman, and were so terrified that they give birth to apes instead of human beings.

Apes have long had a reputation for lacking dignity and morality. Long before Darwin, the essayist Montaigne, chastising human pride, observed in "Apology for Raymond Sebond" that of all animals the apes, "those that most resemble us," were "the ugliest and meanest of the whole herd." Quasimodo, the hero of Victor Hugo's novel *The Hunchback of Notre*

Dame (1831) was certainly based partly on reports of anthropoid apes that were filtering back to Europe during the time Hugo was writing. The character was deformed and could barely speak, yet he had superhuman strength and agility. He climbed like an ape among the gargoyles and demons in the remote corners of the cathedral. His tragedy was to be almost human, yet not quite. He could feel the same passions as other men, yet could not share their lives.

Just as the process of distinguishing apes and men was nearly complete, Darwin developed his Theory of Evolution with *The Origin of Species* in 1859. Not everyone could understand the book, and some thought that Darwin was either blasphemous or crazy. In a famous debate in 1860, Bishop Wilberforce asked Thomas Huxley whether the ape was on his mother's or his father's side of the family. Huxley replied that rather than be descended from a gifted man who mocks scientific discussion, "... I unhesitatingly affirm my preference for the ape." His brilliant rhetoric may have won the day, but wisecracks about apes for grandparents were constantly used in the vitriolic debate about evolution.

In the early twentieth century, racist caricatures, whether of Africans, Jews, Irish, or Japanese, usually showed people slouched over in an ape-like fashion. Adolf Hitler wrote in *My Struggle* that Germans must dedicate the institution of marriage "to bring forth images of the Lord, not abominations that are part man and part ape."

In the early 1980s, experiments in teaching great apes to communicate with human beings, either by computers or by hand signs, generated a great deal of excitement. Jane Goodall and many others observed that apes use tools such as stones to crack nuts or sticks to extract termites from wood. It is a little odd, though, that these observations appeared surprising, since simian use of tools had been regularly noted in natural history books until about the end of the nineteenth century. In 1994, Paola Cavalieri and Peter Singer published *The Great Ape Project: Equality Beyond Humanity*, a book of essays to champion the cause of extending human equality to apes. Few if any of the contributors realized that they were merely reviving a very old debate.

Today, rumors continue to circulate about ape-men such as the yeti. Tabloids announce such exploits as "I was Bigfoot's Love Slave" in supermarkets across the United States and the world. Fantasies of anthropoid apes entertain people in film, from "King Kong" to "Planet of the Apes." Our movies are also full of wild men, from Tarzan of the Apes to Rambo. Throughout the 1990s, middle class American men flocked to "wild man weekends" in the woods, during which they heard lectures and discussed their problems around a fire.

Bigfoot, an ape-like monster up to twelve or thirteen feet high, is regularly reported in the woods of Canada and the United States. Tales of bigfoot originated among Indian tribes of Northern California, and may be traced back to around 1850 in oral traditions. In the early twentieth century, bigfoot was conflated with sasquatch, a similar woodland creature from the legends of Native Americans in British Columbia, Canada. The creature was further conflated with other spirits, bogeys, demons, and specters from other Indian tribes, such as the flying heads and stone giants in the lore of the Algonquians. In the early twenty-first century, there have been hundreds of reports every year by people who claim to have seen bigfoot.

BEAR

So watchful Bruin forms, with plastic care,
Each growing lump, and brings it to a bear.
—ALEXANDER POPE, *The Dunciad*

Of all animals, the bear is probably the one that most clearly resembles human beings in appearance. Even apes cannot stand fully upright, and only walk with difficulty. The bear, however, can run on two legs almost as well as a human. Like a person, a bear looks straight ahead, but the expressions of bears are not easy for us to read. Often the wide eyes of a bear suggest perplexity, making it appear that the bear is a human being whose form has mysteriously been altered. Bears, however, are generally far larger and stronger than people, so they could easily be taken for giants.

Perhaps the most remarkable characteristic of bears is their ability to hibernate and then reemerge at the end of winter, which suggests death and resurrection. In part because bears give birth during hibernation, they have been associated with mother goddesses. The descent into caverns suggests an intimacy with the earth and with vegetation, and bears are also reputed to have special knowledge of herbs. At Drachenloch, in a cave high in the Swiss Alps, skulls of the cave bear have been found facing the entrance in what appears to be a very deliberate arrangement. Some anthropologists believe this is a shrine consecrated to the bear by Neanderthals, which would make it the earliest known place of worship. Others dispute the claim; true or not, the very idea is testimony to the enormous power that the figure of the bear has over the human imagination.

The cult of the bear is widespread, almost universal, among peoples of the Far North, where the bear is both the most powerful predator and the most important food animal. Perhaps the principal example of this cult today is the one followed by the Ainu, the earliest inhabitants of Japan. They traditionally adopt a young bear, raise it as a pet, and then ceremoniously sacrifice the animal. Eskimo legends tell of humans learning to hunt from the polar bear. For the Inuit of Labrador, the polar bear is a form of the Great Spirit, Tuurngasuk. The name of Arthur, the legendary king of Britain, derives from "Artus," which originally meant "bear."

Countless myths and legends reflect an intimacy between human beings and bears. The Koreans, for example, traditionally believe that they are descended from a bear. As the story goes, the tiger and the female bear had watched humans from a distance, and they became curious. As they talked together on a mountainside one day, both decided that they would like to become human. An oracle instructed them to first eat twenty-one cloves of garlic, and then remain in a cave for one month. They both did as instructed, but after a while the tiger became restless and left the cave. The mother bear remained, and at the end of a month she emerged as a beautiful woman. The son of Heaven, Han Woon, fell in love with her and had a child with her, Tan Koon, who is the ancestor of the Koreans.

The Greek deity Artemis, whose name literally means "bear," was the goddess of the moon, the hunt, and animals. The bear was also sacred to

The enormous size of the bear, together with its similarity to human beings, often makes it an object of both awe and derision, and dancing bears were until recently a common feature of traveling shows.

(Courtesy of the Department of Library Services, American Museum of Natural History, #2A4017)

Diana, her Roman equivalent. In a story from the Roman poet Ovid, the god Jupiter disguised himself as Diana, and then raped her nymph companion Callisto. On realizing that Callisto was pregnant, Diana banished the young girl from her presence. Eventually Callisto gave birth to a boy named Arcas. Juno, the wife of Jupiter, turned Callisto into a bear and forced her to roam the forest in perpetual fear. Arcas grew to be a young man. He went hunting in the forest, saw his mother, and raised his bow to shoot her. At that moment, Jupiter looked down, took pity on his former mistress, and brought both mother and son up to Heaven, where they became the constellations of the great and little bear. This is only one version of the story among many, but the Arcadians traditionally trace their origin to Callisto and her son.

The ancient Hebrews, who were herders, regarded carnivorous animals as unclean, and the bear was no exception. In the Bible, the young David protected his flock against bears (1 Samuel 17:34). The bear became a scourge of God when small boys followed the prophet Elisha and made fun of his bald head. Elisha cursed them, and two she-bears came out of the woods and killed the children (2 Kings 2:23-24). According to tradition, however, Elisha was later punished with illness for his deed.

The Tlingit and many other Indian tribes on the northwest coast of the North American continent have told stories of a young woman who was lost in the woods and was befriended by a bear. At first she was afraid, but the bear was kindly and taught her the ways of the forest. Eventually she became his wife. She grew thick hair and hunted like a bear. When the couple had children, she at first tried to teach them the ways of both bears and human beings. Her human family, however, would not accept the marriage, and her brothers killed her husband, whereupon she broke completely with the ways of humans.

Many tales pay tribute to the maternal role of the mother bear. Repeating a bit of lore found in the works of Pliny the Elder and other writers of antiquity, Medieval bestiaries told of cubs that were completely formless at birth. Their mother would mold them with her tongue, literally licking the cubs into shape. The mother bear must constantly protect her offspring from their father, who would eat them out of jealousy and hunger. This fierce protectiveness is part of what has moved contemporary American author Terry Tempest Williams to posit a special bond between women and bears. "We are creatures of paradox," she wrote in an essay entitled "Undressing the Bear." She continued, "Women and bears, two animals that are enormously unpredictable, hence our mystery."

Many European fairy tales suggest such a bond. For example, according to the Norwegian story "East of the Sun and West of the Moon," recorded from oral traditions by George Dasent, a bear went to the house of a poor family and asked the parents for their daughter in marriage, promising great riches in return. The father persuaded his reluctant daughter to agree, and the bear carried her home on its back. The bear visited the young woman every night, but always departed at the break of day, and she was forbidden to know where her husband went. Finally, one night she was over-come with curiosity and lit a candle, only to see him vanish. She made a long and perilous journey to the land east of the sun and west of the moon, where she was finally reunited with her husband. Her love had broken the enchantment of a sorceress, and he turned out to be a human prince. In another version of the tale, three sisters were talking about the men they would marry, when one said in jest, "I will have no husband but the brown bear of Norway." That is indeed what finally happened, but the couple was permanently united only after many trials and tribulations. These stories belong to the cycle of "Beauty and the Beast" fairy tales in which a bride must learn to see past the bestial appearance of her partner in order to find a gentle young man.

One tale that seems to lament the loss of intimacy between bears and human beings is the Icelandic saga "King Hrolf and His Champions," from the thirteenth or early fourteenth century. King Hrolf was drinking with his warriors when the army of Queen Skuld attacked them. Only Bothvar Bjarki, the greatest of the King's knights, was not to be found, and all thought he must have been killed or captured. As the battle raged, an enormous bear appeared at the side of King Hrolf. Weapons simply rebounded off the bear's skin, and it killed more enemies with his paw than any five heroes could have dispatched. One of King Hrolf's champions, Hjalti the Magnanimous, ran back to camp, where he found Bothvar in a tent. Outraged, Hjalti threatened to burn the tent and Bothvar. Calmly and a bit sadly, Bothvar rebuked Hjalti, saying that he had proved his courage many times. He was ready to join the battle, Bothvar explained, but could offer his king more help by remaining behind. Finally, Bothvar reluctantly entered the fray, but the bear disappeared, for Bothvar and the bear were one. King Hrolf and his champions all fought valiantly, yet were overwhelmed by the enormous host of Queen Skuld and killed.

Still another such tale is Valentine and Orson, popular in the Middle Ages, which is preserved for us in French and English texts from around the end of the fourteenth century. The infant Orson was lost in the woods, but a mother bear took him home to her cave and raised him as one of her cubs.

He grew up to be huge, immensely strong, covered with hair, and able to speak only in grunts. For a time, Orson was the terror of the woods, feared by both animals and human beings. When his beloved mother died, Orson let himself be taken by his brother Valentine to the court of King Pepin of France, where he learned the ways of men and became a knight.

A dreaded warrior known as the Green Knight had captured a princess and challenged anyone who wished to rescue her to battle. Many of King Pepin's knights took up the challenge, but the Green Knight bested them all and hung them from a tree. Finally, the turn of Orson came. In the first round of their joust, Orson inflicted several wounds on the Green Knight, but they healed instantly. Realizing the Green Knight could not be defeated by weapons, Orson leapt from his horse, threw away his sword, tore off his armor, pulled the Green Knight from horseback, and forced his adversary to yield, rescuing the princess and winning her for his bride.

In ancient and early Medieval times, Europeans had considered the bear "king of beasts." Several monarchs, particularly in Scandinavia, proudly traced their lineage to bears. The bear, however, had not been integrated into Christian imagery, since that religion had developed primarily in the Mediterranean, where the lion was the apex predator. Gradually, as it became rarer, the lion had gone from being primarily an object of fear and revulsion to being a symbol of Christ, so clergy promoted the "Christian lion" at the expense of the "pagan bear." To accomplish this, bears were mocked by being forced to dance in chains, clumsily imitating a human being in marketplaces for public amusement. Bear-baiting, where a chained bear was set upon by fierce dogs, was also a popular amusement until about start of the nineteenth century. Bears were presented as both cruel brutes and pathetic victims in many stories, and then hunted to near extinction. In the Medieval tales throughout Europe since the end of the eleventh century, "Bruin" the bear is an aristocrat whose natural strength is no match for the peasant cunning of Renard the Fox. But the old awe of the bear continued to resurface in legends and heraldry well into the modern period.

Edward Topsell, the Elizabethan zoologist, reported in 1656 that a man was walking along carrying a large cauldron one autumn day when he saw a bear nibble a root, then descend into a cave. The man was curious and started to chew on the root of the same plant. Immediately he began to feel very sleepy, and he was barely able to throw the cauldron over his body. He remembered no more until he lifted up the cauldron to find the last snow melting on a beautiful spring day.

In William Faulkner's short story "The Bear" (1942) a giant brown bear known as Old Ben becomes the symbol of a vanishing wilderness in the American South. As long as he was there, the land remained wild and strangers did not dare to intrude. The hero of the tale is a boy, learning to be an accomplished woodsman, who found the forests are deprived of both danger and romance after their guardian has been killed by hunters. Like all large meat-eaters, bears had become rare by the twentieth century. The terror that bears once inspired came to be remembered through a haze of nostalgia, and the teddy bear became a favorite toy of children. The name comes from a story that had President "Teddy" Roosevelt, an avid big-game hunter, declining to shoot a bear cub, thinking it unsporting to take advantage of the helpless creature.

No longer greatly feared, the bear has become a symbol of vulnerability. Everybody in the United States who was born before the 1970s or so has seen posters with Smokey the Bear, who was created during World War II to warn people that Japanese shelling might begin a conflagration in the woods of America. When the war ended, the United States Forest Service retained Smokey as a symbol in a campaign to prevent the careless ignition of forest fires. Far from being bestial, he has a rather parental image. He wears human clothes and a forester's hat. His facial expression is mature, friendly, and a little melancholy. Yet if Smokey seems almost absurdly civilized, his role remains that of bears since archaic times—protector of the wild.

BEAVER AND PORCUPINE

*What is the vaunted village of the beaver, the most ingenious of quadrupeds,
compared with a human city, with hits ships and merchandise, its temples,
churches, and dwellings, its libraries, and its monuments of art?*
—S. G. GOODRIDGE, *The Animal Kingdom Illustrated* (1869)

According to a Mediterranean legend, when a hunter chased a beaver,
the animal would bite off its testicles, which contained a medicine known
as "castoreum" (actually in sacs between the pelvis and the base of the tail),
giving the pursuer what he probably wanted and thus escaping alive. This
account was reported by Pliny, Aelian, Horapollo, Cicero, Juvenal, and many
others, and may at times have alluded to the self-castration by priests of the
goddess Cybele in Asia Minor. The tale was later repeated in Medieval bestiaries, where it was interpreted as an allegory of the soul that, pursued by
the Devil, must give up all lewdness.

The beaver, meanwhile, was a popular totem and often a bearer of culture for Native American tribes. According to legends of the Algonquin,
Lenape, Huron, and many other Indians, the beaver first created land (often
helped by the muskrat or otter) by dredging up earth from the bottom of
the sea.

The Blackfoot Indians tell of a man named Apikunni, who had been
temporarily banished from his tribe and took refuge during the winter in a
beaver's house. When he left in the spring, the patriarch of the beaver family
gave him a pointed piece of aspen. Using the stick as a weapon, he became
the first man ever to kill in war, and so he was welcomed back by his people
and made their chief. The Osage tribe sometimes traced its origin to a chief
named Wasbashas, who was taught to build by beavers after he had married
the daughter of their king.

Often associated with the beaver in both Europe and America is the
porcupine, a rodent known primarily for the spikes covering its back. The
most widely spread legend about the porcupine, found in the work of Pliny

This early American print illustrates the indiscriminate slaughter of beaver due to importation of firearms and the demand for beaver pelts. Note, however, the great architectural sophistication of the beaver homes. The animals were reported to "Build like architects," in the words of Oliver Goldsmith.

and many other authors of the ancient world, is that it can shoot its spines when attacked. Aelian added that the porcupine can aim its quills at an attacker with considerable accuracy and that the quills "leap forth as though sped from a bowspring." This fiction is still commonly believed today.

In Native American mythology, the porcupine frequently accompanies the beaver, usually as a companion but occasionally as an adversary. The Haida of the Northwest coast told a story of the war between the clans of Beaver and Porcupine. After Porcupine had stolen Beaver's food, the clan of Beaver placed Porcupine on an island to starve, but the clan of Porcupine rescued their leader when the water froze and they could walk across the ice. The clan of Porcupine then captured Beaver and placed him high in a tree. Beaver could not climb, but he chewed his way down the tree, and the two clans finally made peace. Both are represented by animals that survive by ingenuity rather than power.

In the Romantic period of the eighteenth and nineteenth centuries, a revolt against the constraints of civilization led to the celebration of wild animals. Early explorers were amazed by the size of beaver lodges in the New World. Influenced by the tales of Native Americans, they brought back to Europe fantastic stories of a highly sophisticated beaver society. Beavers were said to build with mortar, use their tails as trowels, and have a system of parliamentary law. By the eighteenth century, the beaver was regularly mentioned—along with the elephant, ape, dog, and dolphin—as perhaps the most intelligent animal after man.

Georges-Louis Leclerc de Buffon, the most popular naturalist of the eighteenth century, believed that all animals once had a civil society with laws, before they were murdered and enslaved by human beings. The last remnants could be found in the New World, where beavers still built villages, created constitutions, and held courts of law. After keeping an imported beaver as a pet, he was disappointed to find it listless and melancholy. He concluded that the beaver did not possess extraordinary native intelligence but merely showed what all animals might be capable of if their social cohesion had not been disrupted by human beings. Oliver Goldsmith, in his enormously popular *History of Animated Nature* (1774), wrote of America,

"The beavers in those distant solitudes are known to build like architects and rule like citizens." He added that the homes of the beavers "exceed the houses of the human inhabitants of the same country both in neatness and in convenience."

But the reputation for intelligence did not protect the beaver, any more than European idealization of the "noble savage" protected the Indians. While many Europeans were romanticizing the American beaver, colonial trappers were finding it a lucrative source of fur. Greed easily prevailed over sentiment, as the British, French, Dutch, and even some Indians engaged in the "Beaver War," an intense competition for furs that often escalated into armed conflict and drove the beavers in North America close to extinction.

The stories of the sophisticated commonwealth of beavers were debunked by the explorer Samuel Hearne—an employee for the Hudson Bay Company, which traded in beaver pelts—in the late eighteenth century. They continued, however, to be repeated in books of natural history, often more interested in entertainment than science, well into the twentieth. Only Native Americans, particularly in tribes of the Northwest Coast, continue to regard the beaver as almost human, or more than human, to this day.

PIG

I am fond of pigs. Dogs look up to us. Cats look down on us. Pigs treat us as equal.

—Attributed to WINSTON CHURCHILL

Human attitudes toward pigs cover an enormous range, but they are consistent in one way. We almost always perceive pigs as being very close to the earth. They are indispensable in looking for truffles in southern France, since their ability to smell things beneath the soil exceeds that of even dogs. They also sometimes take baths in mud to escape the heat. If we see earth as a prison of the spirit, we are likely to hate pigs; if we long for contact with the earth, we may love them. For better or worse, they represent the joys and limitations of the flesh. The pig is holy, yet perfectly at home in Hell.

The pig is gentle, yet harbors such wildness that even devils are terrified. The pig is revered, hated, loved, feared, admired, exploited, laughed at, and regarded as a friend. It is as if the pig were the entire animal kingdom in a single form.

Pigs have a large range of vocalizations, which are very expressive, as well as a keen intelligence. They are often used as substitutes for human begins, in myth as well as medicine. The internal organs of a pig are remarkably similar to those of a human being. This feature helped make pigs a favorite animal for sacrifice to the gods, since the sacrificial victim was generally a symbolic surrogate for a person. Whenever the Roman State entered a contract, a pig would be taken to the temple of Jupiter. As he slit the animal's throat with a sacred sickle, the priest would say, "If the Roman people injure this pact, may Jupiter smite them as I smite this pig." Today, the organs of a pig, especially heart valve replacements, are favorites for transplant into men and women, and pigs are often injected with human genetic material so that their tissues will function after transfer. But this, of course, further blurs the boundary between people and swine, accentuating the ethical problems of sacrificing animals for human health.

Pigs have large litters, which helped make them a symbol of fertility. Ancient Egyptian women who wished to have children sometimes wore amulets depicting a sow and piglets. Pigs would also assist in agriculture by turning over the soil so it could be more easily plowed. Nut, the beloved Egyptian goddess of the sky, was sometimes depicted as a pig. Nevertheless, Set, the evil brother who kills the god Osiris, was also sometimes given porcine form. He is a very early image of the Devil. Traditional devils in Medieval times and even today have the pointed ears and tusks of a boar.

Herodotus wrote that pigs were normally considered so unclean in Egypt that swineherds were banned from temples. Should an Egyptian accidentally touch a pig, he would immediately rush to a river and jump in, not even bothering to undress. Nevertheless, revulsion alternated with reverence. Osiris, the god of the dead, was associated with pigs. At his yearly festival, swine were sacrificed during the night of a full moon. On the next day everyone would eat pork, which was otherwise strictly forbidden. Those

*A monk placing
the papal crown
on a pig.*

*(Woodcut by Hans
Weiditz, Augsburg 1531)*

who were too poor to afford a pig would form one out of dough, which they would then sacrifice.

Several Hindu scriptures including the *Ramayana* contain a story in which the demon Hironyashka tried to hide the Earth, personified as the goddess Bhudevi (whose name means "beloved of the boar"), in primal waters. As she was captured, the Earth uttered a plaintive cry, and the god Vishnu (in some versions, Brahma) came to her rescue. He assumed the form of a boar, Vahara, raised the world above the waters on his tusks, and then killed Hironyashka after a thousand years of combat. The worship of Vahara was suppressed in India under Islamic rule, since Muslims consider the pig unclean, but it remains popular today.

In India, the bloodthirsty goddess Kali was represented as a black sow, perpetually giving birth and eating her offspring in an endless cycle. In Greece, Demeter (Roman "Ceres"), the gentle goddess of agriculture, was also associated with pigs. On the other hand, the Babylonian Tammuz, an agricultural deity, was, like the Greek Adonis, killed by boars while hunting. In the ancient world, boars were often feared not only for their fierceness but also for the damage they could do to crops by eating and tearing up fields. When heroes—such as Meleager, Theseus, and Hercules—killed wild boars, they gained great renown.

In Homer's Odyssey, the sorceress Circe changed the crew of Odysseus into pigs for one year, after which Odysseus forced her to return them to their original form. She symbolizes any temptress who inspires men to behave in a bestial way. Nearly a millennium after the age of Homer, Plutarch wrote a delightful satire entitled "On the Use of Reason by So-called 'Irrational' Animals." In Plutarch's version, at the request of Odysseus, Circe agreed to change the pigs back into men, but only if they themselves wanted to change. She called on a pig named Gryllus to speak for the crew. When Odysseus said that human beings showed greater courage than animals, Gryllus reminded him of the Crommyum sow who, even without the use of weapons, almost defeated the hero Theseus. When Odysseus said that humans showed greater reason, Gryllus gave many examples of animal intelligence; pigs, for example, went to riverbeds and ate crabs to cure their illnesses. The debate ended abruptly and the manuscript may never have been finished, but Odysseus was so completely beaten that it is very hard to imagine a recovery.

In Norse mythology, the god Frey rode in a chariot drawn by the boar Gollinborsti, whose name means "golden tusks." The boar's head, traditionally served in England at Christmastime, was originally a sacrifice to Frey. The boar Saehrimnir had been killed every evening and served to the heroes in Valhalla, to be reborn the next day. In a similar way, pigs in Celtic legend were the food of the gods in otherworldly feasts. The pigs of Manannán, the Irish god of the sea, would also reappear after being eaten.

For the Hebrews, however, pigs were not merely "unclean"; they were the most repulsive of animals. Perhaps one reason was that pigs were carriers of the disease trichinosis, but just about every domesticated animal was capable of spreading some disease. Another possible reason was that pigs had been associated with many pagan mother-goddesses, divinities that the Hebrews abhorred. It could also be because pigs would eat just about anything, while the Hebrews were very fastidious about their food.

The Old Testament tells how the Greek emperor Antiochus Epiphanes tried to force Jews to eat pork, which helped set off the revolt

Various pigs from a nineteenth-century book of natural history.

of Judas Maccabeus. For the most part, Christians initially shared the Hebrew view of pigs as unclean. Matthew told us not to "cast pearls before swine" (2 Macc. 7:6). In Mark, when Jesus cast out demons from a madman, they entered a herd of swine, which then ran out into the sea and were drowned.

To later Christians, the Jewish avoidance of pork appeared to be something like a taboo against cannibalism, and at times even worship of the pig.

During the first century CE, Petronius Arbiter wrote in a poetic fragment, "The Jew may worship his pig-god." The Spanish Inquisition tested Jews who claimed to have converted to Christianity by requiring them to eat the meat of pigs. In numerous popular stories, Jews were turned into pigs. A chronicle of wonders published in Binzwangen, Germany, in 1575 reported that a Jewish woman gave birth to two piglets. At the end of the Middle Ages in Europe, a popular anti-Semitic motif was "the Jew's sow," an enormous pig suckling Jewish men.

Medieval Europeans considered the pig to be the animal closest to human beings. Pigs were raised in a way that straddled not only the boundary between person and animal, but also that between pet and livestock. A pig would be allowed to run freely, fed scraps from the family table, nurtured primarily by women, and, for many purposes, regarded as part of a human family. Then, around the time of the winter solstice, after the members of a family had all paid their last respects, the pig would be ceremoniously slaughtered. This would be followed by the Feast of Saint Pig, during which feasting, games, singing, dancing, and plays would last throughout the night. The bones and inedible parts of the pig would later be ritually buried, in expectation of resurrection.

A pig, usually immaculate, was often painted alongside the hermit Saint Anthony. An Italian tale from the Mediterranean islands, retold by Italo Calvino, starts at a time when all fire was in Hell, and so no hearths warmed families in winter. People, shivering so badly that they could barely speak, appealed to Saint Anthony for help. The holy man went down to the very gate of Hell and knocked with his staff. At his side, as always, was his faithful pig. A devil opened the door a crack, looked out, and said, "Get out of here! We know you. You're a saint. Only sinners are allowed in Hell!" The pig would not take "no" for an answer, forced open the door, knocked down the devil, scattered a pile of pitchforks, and raised so much hell in Hell that the devils were terrified. "Come in and get your pig!" shouted the devils to Saint Anthony. The saint walked in and touched the pig lightly with his staff, at which point the animal became completely calm. "Now get out of here, both of you, and don't ever come back again!" shouted the devils. Without a word or even a grunt, Saint Anthony and his pig walked away. What the devils didn't know was that Saint Anthony was carrying a spark of fire concealed within his staff. As soon as Saint Anthony and his pig reached the surface of earth, the holy man swung the staff above his head so that sparks flew in all directions. And so, thanks to Saint Anthony and his pig, people can tell stories in comfort around the fireplace while the ground is covered with snow.

But the near human status of pigs in the European Middle Ages could bring expectations of human behavior, often of adherence to human laws. In 1389, in the town of Falaise in Normandy, a sow was judicially condemned for killing an infant after a trial lasting nine days. As punishment, the pig, dressed in human clothing, was tortured and then hanged by the neck until dead in front of a crowd consisting of local gentry, peasants, and many other pigs brought in from the surrounding area. There were scores of such animal trials in Europe during the Early Modern Period, and the overwhelming number of them were of swine.

The fact that pigs foraged freely in woods and were even allowed to enter homes made them particularly vulnerable to judicial indictments. Prosecutors sometimes alleged that pigs had an infernal smell, showing their association with the Devil. Their grunts and squeals, which seemed disrespectful to the courts, made things even worse. Plenty of pigs were convicted, and some acquitted, in courts throughout Europe for such offenses as eating their own young or having sex with human beings. Those found guilty were usually either hanged or burned alive.

Aristocratic families of Medieval Europe largely took their animal symbolism from warrior religions of their pagan past rather than Christianity. The nobles admired boars for their military virtues, and boars were among the most popular animals in heraldry. When hunted, boars charge and fight to the end no matter how many dogs and men they face. Social position in Medieval times determined which animals one was allowed to hunt. As a noble animal, the boar was second only to the stag in status, and it presented an even greater test of a hunter's skill and bravery. Metaphors for love were often drawn from hunting. In the late Medieval British ro-

mance *Sir Gawain and the Green Knight*, the wife of his host tried to seduce
Gawain several times. Once, when Gawain resolutely rejected her seductions, he was compared to a boar confronting a hunting party directly and
without fear.

Since domestic pigs were allowed to roam relatively freely until
around the start of the modern era, they would often interbreed with wild
boars. Until the nineteenth century, they still had gray hair and tusks. What
we usually think of when we think of as "pigs"—pink, almost hairless, and
very fat—appeared only fairly recently, having been created from albino varieties. The physical change brought many modifications in both the use
and the symbolism of pigs. They became an image of those spoiled by the
comforts and privileges of civilization.

The boar is the last cycle of the Chinese zodiac, and those who are
born in the year of the boar are said to be courageous but stubborn. The
domestic pig in Asia shared a reputation with its Western counterparts for
appetite and earthy charm. The sixteenth-century Chinese epic *Journey to
the West* told of a monk who undertook a pilgrimage from China to India
to bring back Buddhist scriptures and save China from chaos. The animals
that accompanied him included a monkey, a horse, a sea monster, and Old
Hog, a pig that subdued demons with his rake. Old Hog may have been a
formidable fighter, but laziness or appetite easily overcame him. As a reward
for his good services, he was finally made not a Buddha but Janitor of the
Altars, and he had the pleasant task of eating scraps left after celebrations.

In Berlin during the 1920s, there were several riots by veterans, working men, Nazis, and Communists. The police who were summoned to put
down the violence were called "pigs." In Nazi Germany, Minister of Agriculture R. Walter Darré wished to proclaim the pig the central animal of
the Aryan people, but other Nazis identified pigs with Jews. Student rebels
throughout much of the world took up this epithet again in the 1960s,
taunting both politicians and law-enforcement officers by calling them
"pigs." In 1968, protesters at the Democratic convention in Chicago held a
mock convention and nominated a pig for president.

In his novella *Animal Farm* (1946), George Orwell used the modern

Domestic pigs are generally looked on with affection in books for children, in part because of their dependence and vulnerability.

{Illustration from the early twentieth century by W. Heath)

farm as an allegory for the totalitarian state. Pigs, as the most intelligent of animals, lead a revolt against the brutal farmer Jones. "All animals are equal . . . " the pigs proclaim, but they later add, "some are more equal than others." A Berkshire boar named Napoleon drives out his porcine rivals, learns to walk on two legs, and exploits the other animals as much as human beings had ever done. The very last sentence of the book is, "The creatures outside looked from pig to man, and from man to pig, and from pig to man again; but already it was impossible to say which was which."

That depiction of pigs may be good literature, but it is still rather ungracious. Pigs are among the most useful of animals to human beings. Just about every part of the body of a pig is used; pudding is made from the blood of pigs, and sausages are wrapped in the intestines; the leather of a pig's skin is highly prized. The ability of pigs to digest almost anything and convert it into edible material makes them especially helpful to farmers. Many pigs receive remarkably little gratitude, and they are often kept in cramped, filthy conditions until the time of slaughter arrives.

Pigs are still among the most beloved figures in books and movies for children. These include Wilbur (from E. B. White's *Charlotte's Web*), Porky Pig, and Babe, all gentle figures who show little of either the valor or the filthy habits traditionally associated with swine. Miss Piggy, one of the biggest stars of *The Muppet Show*, is a modern heiress to ancient porcine goddesses such as Nut. She has starred in feature films, written a popular book on fashion, and been featured on posters and calendars. Miss Piggy is forever flirting. She may act vain and clumsy, but you had better not laugh at her too openly. She has the superhuman strength and fierceness of her porcine ancestors, which so impressed people in ancient times. The pig, in summary, is the only animal in the Western world that not only transcends the division between wild and domestic animals but even that between pets and livestock.

2
TRICKSTERS

TRICKSTERS, OFTEN ANIMALS, ARE DEFINED FAR LESS BY TANGIBLE qualities than by a very fundamental ambiguity. They are frequently shape-shifters, and they almost always try to prevail by wit and deception rather than by brute force. Because of their equivocal nature, they frequently double as culture heroes and as clowns. Among the important anthropomorphic tricksters are Prometheus and Dionysius in Greek mythology, and Loki in Norse. There are innumerable tricksters in the more zoomorphic mythologies, especially those of Africa and the Americas, including Anansi the spider among the Ashanti people, Mishaabooz the Hare among the Algonquin Indians, and many more.

Since the trickster figure was first identified as a cross-cultural phenomenon by folklorist Daniel Brinton around the end of the nineteenth century, academics have examined it intensely. One theory holds that such figures belong to a stage of cultural evolution when individual personalities were not yet clearly differentiated, while another maintains that the trickster is an archetype in the human psyche. It is also possible, however, that tricksters only seem incoherent because scholars do not fully understand their societies. The blending of apparently divergent personalities involves a delicate balance, which may be extremely difficult to comprehend across lines of culture. There is a strong temptation to simplify these figures, viewing the trickster one-dimensionally in terms of a quality such as cleverness, ambition, obscenity, or service to humankind.

The raven, particularly among Indians of the Northwest Pacific Coast might have been included here, but, while often comical, its cosmic power may place it in a more exalted category. It will be found in Chapter Three on "animal sages."

COYOTE, FOX, AND JACKAL

One can gloss, think, study, and muse
More upon Renard than anything else in the world.

—Anonymous, "The Romance of Renard"

The fox and jackal are predators of moderate size, a trait that has probably made them easier for most people to identify with them rather than with the huge lion or the ferocious wolf. The two are virtually inter-changeable in Near Eastern literature. Both of these canids are renowned for their cleverness. The "sagacity" of animals in folklore is often a rationalization of magic, and archaic manuscripts confirm that these animals once appeared as powerful sorcerers. The coyote is a canid of about the same size as the fox and jackal, but it is indigenous to the New World. In a striking instance of the universality of animal symbolism, it has much the same role in Native American lore as that of its cousins in Eurasia.

In one of the very earliest surviving literary manuscripts, written in Mesopotamia about the middle of the third millennium BCE, a fox brings back Enki—son of the god of air, Enlil—from the Netherworld. Enki, this Sumero-Babylonian god of magic, continued to be associated with a fox in Sumerian and Babylonian mythology. In another cuneiform manuscript, Enki had disobeyed Ninhursag, the great earth mother, and she punished him with the curse of death. The other gods gazed on helplessly as Enki sank into oblivion, when the fox appeared and brought the deity back. The tale may have originated with a shamanic trance, in which Enki entered the realm of the dead as his body was possessed by a fox.

By the second millennium the role of the fox as a trickster was already established in Mesopotamian animal proverbs, the ancestors of Aesop's

Frontispiece by Paul Meyerheim to a ninenteenth-century collection of tales about Renard the Fox.

fables. A Babylonian tablet entitled "The Fable of the Fox" from the early first millennium BCE told of a fox, wolf, and dog that brought suit against one another before a lion. They accused one another of sorcery, theft, and, most especially, of provoking the gods to bring a terrible drought that threatened to destroy the world. Much of the manuscript is missing, but the fox seemed to carry the day with his cleverness. In the final tablet, we learn that rain had come and the fox was entering the temple in triumph.

The Egyptian equivalent of this magical fox was the god Anubis, depicted with a human body and the head of a dog or a jackal. Anubis is a psychopomp, who guides the dead to their place of judgment, and then weighs the heart of the deceased against Maat, the spirit of cosmic order, which is often represented by an ostrich feather. If the heart sinks on the scales, the deceased will be devoured by demons, but if the heart rises he or she might join the god Ra and sail across the sky in the boat of the sun. That jackals burrow may have suggested intimacy with the earth, while their scavenging kills may also have contributed to an association with the dead.

The fox was a trickster in the fables attributed to the legendary Aesop in Greco-Roman civilization. He constantly matched wits with other animals, though he was generally obsequious to the lion. The most famous of these tales was known as "The Fox and the Grapes," and the story could hardly be simpler. A fox looked up at grapes on a trellis. He tried repeatedly to reach them by jumping without success. Finally, the fox said, "They are probably sour anyway" and walked away. This anecdote has been told in various eras with different morals. In Medieval versions, the fox is called wise, while in modern ones he is mocked as foolish. For a trickster, even a frustrated one, wisdom and foolishness are often very close indeed.

The Bible, however, took a less anthropomorphic view of animals, which were often credited with pathos but rarely with wit. Foxes were associated with trickery, but it was as hapless implements rather than perpetrators. Samson caught 300 foxes (or, possibly, jackals), tied them in pairs by their tails, fastened a torch to each pair, and set them loose in the cornfields of the Philistines (Judges 15:4). The fox received increasing attention and respect in Jewish literature of the Diaspora, where political sensitivities forced leaders to express themselves indirectly by means of fables and parables.

Rabbi Akiba once defied Roman authorities by teaching the Torah. When a follower asked him if he was afraid of the government, he replied with a story. A fox once asked some fish why they kept moving from one place to another, and they replied they were fleeing the nets of the fishermen. The fox invited the fish to come out on dry land and live with them in peace. A wise fish replied that the danger they faced in their own element must be much less than what they would face in a foreign one. In a similar way, the Jews would face greater danger if they abandoned their traditions.

Illustration by Richard Heighway to Aesop's fable of "The Fox and the Mask."

The Talmud also contains many stories that celebrate the wit and wisdom of the fox. Without their own army or police, the Jews of the Diaspora had to live by wit and diplomacy, and they often identified with the crafty fox in respect to both his virtues and failings.

The dual nature of the jackal and fox is especially clear in the Hindu-Persian *Panchatantra*, also known *as The Fables of Bidpai*, which was probably written down in about the second or third century BCE. The frame of this collection of stories concerned two jackals, the devious Damanaka and the honest Karataka, at the court of the lion king. Damanaka becomes jealous when the lion adopts the ox Sanjivaka as his favorite. Against the counsel of his companion, the insidious jackal stirs up strife between the lion and ox by telling each lies about the other. Finally Damanaka provokes the former friends to engage in a battle, in which the lion is wounded and the ox is slain. The debates and intrigues provide occasions for all of the various characters to tell stories in support of their advice.

With trickster figures, excessive cleverness is often a form of folly. In one of the tale of Bidpai, a jackal strayed into the city where he was chased by dogs. In desperation, he jumped into an enormous vat of blue dye to hide. When he could no longer hear any barking, the jackal slowly climbed out, and returned to the jungle. When the other creatures saw this strange blue beast passing, they were terrified and thought he possessed supernatural powers. The jackal had himself declared king and made the lion his Prime Minister, and the elephant and monkey became his royal attendants. The jackal then assigned some role in its kingdom to every creature except for the other jackals, which he banished out of fear that they might expose him. One day, however, the blue jackal heard other jackals howling in the distance. Unable to restrain himself, he began to howl with them. Realizing what had happened, the beasts were outraged and tore the blue jackal to pieces.

The *Panchantra* was translated into Arabic by Mohammed al Haq, and then it became known as *Kalila wa Dimna*. The Arab version, in turn, was translated into Latin and eventually into all major European languages during the late Middle Ages and Renaissance. It provided the foundation for the cycle of Renard the Fox stories, which were first written down in French toward the end of the twelfth century and quickly spread throughout Europe. The fox here is the clever peasant, who outwits the other animals, especially his more powerful, aristocratic adversaries such as the wolf and bear. His opponents end up ruthlessly beaten, cuckolded, maimed, eaten, and otherwise destroyed. Any sympathy we may have for the fox as the "underdog" is at least severely tested by his unscrupulousness. Depending in part on the class affiliation of the authors of various manuscripts, Renard comes across as a thorough villain, a flawed hero, or simply a figure of raucous fun. In one popular story, the lion was ailing, and the fox persuaded the king of beasts that the cure was to wrap himself in the skin of the wolf. The wolf, Renard's great adversary, accordingly was flayed, but the lion died soon after, leaving Renard alone in triumph. Such sophisticated authors as Chaucer, La Fontaine, and Goethe later retold folktales of Renard.

Readers will find a complex and thoughtful perspective in the *Fox Fables* of the French Rabbi Berechiah ben Natronai ha-Nakdan, sometimes known as "the Jewish Aesop," which were written in Hebrew around the end of the twelfth century. These fables are a bit reminiscent of the Old Testament, filled with violence and iniquity yet told with a highly moralistic gloss. Although most of the fables do not center on the fox, the author ~~uses~~ sometimes uses the fox as a spokesman. Much as the Jews often had to look to monarchs for protection, the fox of these stories has to cultivate a relationship with the lion-king. The fox of ha-Nakdan is highly pragmatic but, unlike that of the Renard cycle, rarely vicious. In one fable, the fox gnawed the bones of goats killed by the lion. When the lion rebuked him, the fox said he was ashamed and promised not to steal from the lion again. Ha-Nakdan concludes with a moral that we should, like the lion, forgive those who wrong us.

In the folklore of the Merovingians, who ruled Germany and much of France at the beginning of the Middle Ages, the creatures of the forest have a court. The lion is king, while the bear and stag are nobles. Every year at the summer solstice they meet to hear lawsuits and dispense justice. As human rulers in Europe appeared less glamorous and more corrupt, people also took a more jaundiced view of the animal kingdom. In stories of Reynard the Fox, told throughout Europe toward the end of the Middle Ages, the king of beasts was just a fool for all his pomp. The wolf and fox quarreled over chickens stolen from the coop in the language of piety and romance.

As the peasants were gradually emancipated from serfdom and the Jews from the ghettos, the traditional elites sometimes took out their residual class anger on the fox. As the hunt of stags and boars ceased to be an aristocratic privilege, the nobility, especially in England, would pursue the fox instead. Through the Middle Ages, the fox had been considered unworthy of being hunted by a lord. During the modern period, the foxhunt became a ritualistic affirmation of the feudal order, conducted in the most ceremonious way with elaborate calls, rituals, and uniforms. Modern sportsmen such as John Mansfield and Siegfried Sassoon wrote passionate accounts of the foxhunt, and prints of the chase became popular on living room walls. But many people were also distressed by the spectacle of so many men, ladies, children, dogs, and horses all arrayed against one diminutive creature, and the foxhunt was banned in all of Britain by 2004.

In East Asia, the fox of folklore has remained associated less with wit than with magic, and has generally been female rather than male. Foxes in

Asia are symbols of marital fidelity, and the vixen is also an icon of maternal love. Early Chinese writings on the sanctity of marriage often invoked the model of the vixen in urging mothers not to practice female infanticide. But in their dealings with human beings, foxes are not necessarily bound by the same codes or loyalties. As shapeshifters, foxes often assume the form of beautiful women in order to seduce men. They frequently draw the life force from men, yet sometimes they also truly fall in love with their partners. While they may fool human beings, fox maidens are sometimes recognized by other animals. They can also be identified by placing a mirror in front of them, because they either do not leave any reflection in a mirror or else show their true vulpine countenances.

Among the earliest recorded stories about these shape shifters is "Jenshih, the Fox Lady," written in China by Shen Chi-chi around the end of the seventh century CE. A poor soldier named Cheng Liu saw a lovely lady walking through the streets and gallantly offered her his donkey as a mount. The pair fell in love, and one day Jenshih confessed to him that she was really a fox, but she offered to remain with him in human form if he would not reject her on that account. When Cheng Liu accepted the offer, she not only proved to be a loving wife but also brought her husband prosperity by managing his affairs with tact and skill. One day, at the marketplace, however, some dogs caught scent of her. Jenshih immediately fell to the ground, assumed the form of a vixen, and began to run. Cheng Liu followed as best he could but was unable to save her from the hounds. More than anything else, the fox maiden in Asia represents the female realm, enticing and frightening men through its secrets.

In Japanese legends, magical foxes or "kitsune" have a society that parallels that of human beings. They have supernatural abilities that just about balance human technology, and they are curious about men and women, just as we are about them. They can easily produce the sort of things that people value such as gold and jewels, but they attach no significance at all to these. Instead, they value things that people find insignificant, such as a pile of straw to make a home in. When people befriend foxes, they will often receive good fortune in return. At times, kitsune assume human form,

Foxes are said to gather annually at Oji near Tokyo. A large number of foxes at the congregtion, according to legend, means a plentiful harvest the next year. The flames emanating from the foxes are known to Japanese as "kitsune-bi" or "foxfire."
(*The Metropolitan Museum of Art, #MMA60544B*)

usually that of a young maiden. They are every bit as individual as human beings are, and some kitsune will seduce men, in the shape of beautiful women, in order to destroy them, but others make the most devoted wives. If a young fellow sees a beautiful woman alone in the woods, she is probably a kitsune; if he approaches her, he will be taking a big risk. Among the few unequivocally benign foxes in Asian lore are the messengers of Inari, the Japanese god of rice, who himself is often depicted as a fox.

Much like Renard the fox, the coyote of folklore is both a sage and a buffoon. The greatest difference perhaps is that Coyote of Native American legend is far more of a cosmic figure than any of foxes or jackals of folklore. In a story told by the Zuni Indians, Coyote and Eagle once stole a box containing the sun and moon from the world of spirits, so that there would be light for the world, but Coyote opened the box out of curiosity. The heavenly bodies then flew away, thus there is winter as well as summer. The Klamath tell how Coyote won fire from Thunderer by cheating at dice. The Pueblo Indians recount that Coyote once helped a great magician to make human beings of clay, and then brought them to life by baking them in an oven. Unfortunately, most people were flawed because Coyote took them out of the oven either too early or too late. Coyote is also widely credited with bringing death into the world. It should surprise nobody that the European legends of Renard and the Native American myths of Coyote have blended in the tales told in the pueblos of Mexico and the American southwest.

Capitalist society always loves tricksters, and so Coyote remains very popular today. Some Indians have complained that white interpreters of their traditions emphasize the undignified and amoral aspects of Coyote at the expense of his holy qualities. One inheritor of that tradition is Wile E. Coyote, a cartoon character who entertains people with his fanatic but usually futile pursuit of a bird called Road Runner. Wile often ends up falling off a cliff or being run over by a car. At first, he might appear to be too much of a loser to be a successor to the Native American trickster. On the other hand, whether Coyote won or lost was always less important than his continuous survival, which is something that Wile does remarkably well.

HARE AND RABBIT

> *What is it—a hopper of ditches, a cutter of corn, a little brown cow without any horns? Answer: A hare.*
>
> —Irish Riddle

The rabbit and hare, members of the family Leporidae, are rodents, yet they usually have a far more benign reputation in folklore than their relative the rat. While rabbits are highly social, hares often tend to be solitary. They are also larger than rabbits and have smaller litters. Nevertheless, the two are often conflated in folklore, and often the same stories are told of both. Like many other prominent animals in myth and legend, leporids are dramatically distinguished from other animals by a single feature—their long ears. Though not terribly fast runners, they are remarkably agile, and their ability to elude predators by changing direction instantly has contributed to their reputation as tricksters. Even more than other rodents, they reproduce prolifically, which has made leporids, especially rabbits, symbols of fertility throughout the world. They are endearingly timid, arousing an affection that often makes it hard for farmers to shoot them, even when they ravish planted fields. The most remarkable feature of their lore, however, is the widespread notion that a hare may be seen in the moon, which is found in many cultures across the globe including those of the Chinese, the Khoikhoi (also known as "Hotten-tots," a term some find derogatory), and the Mayans.

The widespread association of the hare with the moon cannot simply be due to the contours of lunar landscapes, since people envisage the hare in the moon in very divergent ways. Part of the reason is that the leaps of a hare suggest the rising moon, and the patterns of white, gray, and brown on the bodies of hares are suggestive of the lunar surface. The most important reason is probably the extreme watchfulness of hares, which stand almost completely still with their eyes wide open and their ears raised. In an analogous way, the moon, especially when full, which resembles an enormous eye that continually watches events on earth.

The *Jakatas*, early Buddhist animal fables from India centered on the previous lives of Buddha. Once the future Buddha was a hare and lived together with three wise animals, a monkey, a jackal, and an otter. He preached to the other creatures of the forest, telling them not to refuse alms. Sakra, the god of thunder, heard him, and came down to the forest in the guise of a Brahman. The monkey offered him fruit, then the jackal offered meat, and the otter offered fish. Finally, the Brahman came to the hare, which directed him to gather wood and start a fire. When the fire was blazing, the hare hopped in, for he had resolved to offer his own body as food. The flames, however, would not burn. The Brahman revealed that he was truly a god. Then he squeezed a mountain to make ink, and he drew an image of the hare in the moon.

The Chinese see a hare with a mortar and pestle, grinding the elixir of life, in the moon, an idea inspired by the reproductive powers of the animal. This comes from the story of Chang-O, the beautiful wife of the famous archer King Ho-Yi. At a festival, the Queen-Mother of Heaven gave Ho-Yi a pill containing the elixir of immortality. Having drunk much wine, the King wished to sleep before taking the pill, so he entrusted it to Chang-O. She swallowed it herself, immediately felt very light, and then discovered that she could fly. On awakening, Ho-Yi demanded the pill, and, when Chang-O failed to produce it, he threatened to kill her. Chang-O flew away to the moon, where she hid in a cave. One day she coughed up the pill, which immediately changed into a white hare. Chang-O demanded that the hare recreate the pill, and she gave the hare the tools of an alchemist, particularly a mortar and pestle, for that purpose. Being once more subject to the ravages of age, she turned into a three-legged toad as she waited for the hare to finish. In the moon, you can see Chang-O, both as a woman and a toad, standing before a cassia tree and watching the hare, with a pestle between its paws, pounding out the elixir of immortality on a mortar.

A story in the third book of the Hindu-Persian *Panchatantra* may affectionately poke fun at the association between the leporids and the moon, though it also shows the rabbit (in some versions, a hare) in the familiar role as a trickster. A herd of elephants discovered the paradisial Lake of the Moon, and, in their eagerness to drink, they crushed many rabbits to death. A rabbit named Victory went next day to the king of the elephants, saying he was an envoy from the Moon and protected by the laws of diplomacy. He rebuked the King and his herd for killing rabbits, which had been under the protection of the Moon, and he spoke so eloquently that the Monarch decided find the moon and beg forgiveness. Victory lead the King to a place where the full moon shone brilliantly in the water of a lake. When the elephant tried to bow down before it, his trunk touched the water, breaking the image into thousands of pieces. At that Victory said, "Woe, woe to you, O King! You have doubly enraged the moon." The elephant then promised never to return, and the rabbits once again had the lake to themselves.

The fable of "The Tortoise and the Hare."

(IIllustration by J. J. Grandville from Fables de la Fontaine, *1839)*

In the fables of Aesop, the hare is also a trickster, though, like most other tricksters, he often becomes the victim of his own cleverness. Perhaps the most famous fable of all is "The Tortoise and the Hare." The hare had mocked the slowness of a tortoise, which then challenged him to a race. The hare agreed, and, as the two began, he spurted ahead, gaining a great lead. Supremely confident, the hare dawdled, rested, and played, until the slow but steady tortoise overtook him to claim victory.

Caesar states that the hare was sacred to the early Britons. According to the Roman historian Dio Cassius, Queen Boadicea, who led the Britons in revolt against Roman rule, would release a hare from the folds of her dress before each campaign. The direction in which the hare ran would be used to predict the outcome of the battle. The Easter Bunny, originally a hare, was probably a sacrificial animal, offered to the gods at the start of spring. Closely associated with the Easter Bunny are the colored eggs, which go back to pre-Christian celebrations of spring in Slavic lands. At one time eggs may have accompanied the hare not only in games but also in a festive meal.

In the European Middle Ages, hares were often a form in which witches ran about at night. One confessed Scottish witch, Isabel Gowdie, claimed that she had taken the form of a hare when hounds surprised her. She managed to evade them by running into a house and hiding long enough to say the rhyme which disenchanted her, though she still carried a mark on her back where a hound had nipped at her. In *Precious Bane* by Mary Webb (1924) a novel set in the countryside of early nineteenth century Shropshire, the heroine, Prudence Sarn, has a harelip, a slit upper lip like that of a hare. Her mother thought that a hare running across her path shortly before Prudence was born had caused the deformity. Fellow villagers constantly suspected the young woman of a connection with the Devil.

All across Africa, the hare is an important trickster figure, and he often matches his cleverness against the size and strength of a hyena or a lion. In one Hausa story from Nigeria, the lion had so terrified the other animals of the forest that they made a deal with him. One animal was to voluntarily sacrifice himself to the lion every day, and, in return, the king of beasts would refrain from hunting. After a gazelle, an antelope, and many other animals had given their lives, it was the hare's turn. The hare told the lion that he had brought honey as a special present, but another lion, which was even fiercer than the first, had demanded the gift. When the king of beasts demanded to know where his challenger was, the hare pointed to a well. The lion looked into the well, saw his own reflection, pounced, and was drowned. All animals acclaimed the hare as the new king of beasts.

Master Rabbit is a trickster among several Native American tribes. In an Ute Indian tale, Rabbit had become angry because the sun burned his back. He tested his skills by killing all the people and animals that crossed his path, until he felt mighty enough to duel the sun itself. When he hurled a magic ball at the sun, fire spread all over the earth, and Rabbit was seared so badly that he began to cry. When his tears had finally put out the flames, Rabbit realized that killing does not solve problems.

The trickster hare has many names in Indian lore. Among the Ojibwa he is known as "Nanabozho," and among many Algonquin peoples he is "Mishaabooz." In the lore of the Lenape Indians, he is "Tschimammus," one of the twins born to the earth mother after she had fallen from the

Brer Fox laughs uproariously as Brer Rabbit gets stuck to the Tar Baby.

Illustration by A. B. Frost to an Uncle Remus tale by Joel Chandler Harris)

clouds. He ascended to heaven, and, since he was expected to return to earth, Indian converts to Christianity identified Hare with Jesus.

But perhaps the best known, and certainly the most controversial, leporid of modern times is Brer Rabbit, from the late nineteenth century tales that Joel Chandler Harris put in the mouth of an old black slave named Uncle Remus. Brer Rabbit is as ruthless as he is clever, and he continually matches wits with larger predators such as the fox, wolf, and bear. His adversaries can end up not only defeated but also humiliated, cuckolded, roasted, or otherwise horribly punished, though Brer Rabbit himself is also often outwitted by other tricksters such as Brer Terrapin and Mr. Buzzard.

The tales are as controversial as they are popular. While some critics admire the cleverness of Brer Rabbit, others consider him a racist caricature of blacks as shiftless and amoral, but the controversy is probably misplaced. Most folklorists now agree that Brer Rabbit is a version of an Algonquin trickster hare, which entered African-American folklore through Indians who were enslaved along with blacks. The farcical plots make political moralizing seem rather heavy handed, and, for all his wit, Brer Rabbit, like most of the other animals in the tales, is a model of how not to act. But despite an Indian origin, the Brer Rabbit tales remain an important part of African-American folklore, and those who believe the versions of Harris are too preposterous, violent, or patronizing may prefer those of Zora Neale Hurston and others.

The most famous story told by Harris is that of Brer Rabbit and the Tar Baby. Brer Fox made a little figure out of tar, left it in the bushes, and watched until Brer Rabbit came down the road. When the Tar Baby failed to return his greeting, Brer Rabbit became angry and punched the figure. His paw stuck to the tar, so Brer Rabbit struck again and again, until finally all his limbs were bound together by the pitch. Brer Fox had the culprit completely at his mercy, and was trying to choose the most dreadful punishment, when the clever rabbit begged not to be thrown into the briar patch. Brer Fox obliged by tossing the rabbit straight into the brambles. "Bred and bawn in a briar patch" shouted Brer Rabbit as he ran away.

In the nineteenth century, rabbits and hares became favorite figures in books for children. Peter Rabbit, created by Beatrix Potter, is perhaps the most beloved, but just about everybody is also familiar with the White Rabbit and the March Hare from Lewis Carroll's *Alice in Wonderland*. Potter makes the traditional trickery of leporids in folklore into childish misbehavior, while Lewis Caroll makes it into what, from a child's point of view, is the insanity of adults. Far less appealing, yet in some ways closer to folk traditions, is Bugs Bunny, a cartoon character created by Warner Brothers in the 1940s. An amoral trickster, Bugs outwits the dim-witted hunter Elmer Fudd, who often ends up falling from a cliff or being run over by a truck. In the 1970s Richard Adams tried to give greater dignity to rabbits in his novel *Watership Down*, about a group of male rabbits who set out to found a new warren. Although rabbits and hares are not patriarchal, virtually all of these popular images are male.

In *Playboy Magazine*, young girls are referred to as "bunnies." At parties they put on skimpy costumes with long ears and white tails, and every issue features a "bunny of the month." These practices build on the reputation of rabbits for being cute and cuddly. Such use of fertility symbolism is paradoxical, however, since the male clientele supposedly wish to remain unattached, and are certainly not looking at the "bunnies" as potential mothers for their children.

SPIDER

> *The Soul, reaching, throwing out for love,*
> *As the spider, from some little promontory, throwing out filament*
> *after filament, tirelessly out of itself, that one at least may catch*
> *and form a link, a bridge,*
> *a connection*

—WALT WHITMAN

The spider is an image of fate in its relentlessness, as well as in its combination of terror and beauty. The image of a spiderweb gleaming in the dew recalls the stars against the Milky Way. Furthermore, spiders certainly have abilities that could move even a goddess to envy. Even today, engineers have not managed to create a filament with the same combination of thinness, flexibility, and tensile strength as that of a spider's web.

Nevertheless, the usual manner in which many spiders devour their prey can make anybody shudder. When a fly is caught in a web, the spider will inject it with digestive juices and go away, returning later to eat the prey a little at a time. Among certain species, particularly the garden spiders of southern Europe, the female will at times devour the male upon mating, an eerily literal expression of the primal unity of conception and death. Spiders usually have eight eyes, enabling them to see in almost every direction. Nobody can meet, much less read, the gaze of a spider. Two eyes are so much the rule among animals from whales to grasshoppers that any other number is viewed as grotesque. Folklore constantly exaggerates the fearsome attributes of spiders, especially the deadliness of their poisons.

Ovid traced the origin of spiders to the story of Arachne, a young girl who was so skilled at spinning and weaving that even the nymphs gazed on her with wonder. She had boasted that her skill exceeded even that of the goddess Athena. Upon hearing this, the goddess took on the shape of an old woman and went to Arachne, warning her against arrogance. When Arachne refused to retract her boast, Athena revealed herself and challenged Arachne to a weaving contest. Even then, Arachne was not intimidated, and she accepted without hesitation. On her loom, Athena wove pictures of mortals who had dared to measure themselves against the divinities and met their doom. On her loom, Arachne wove pictures showing the follies of gods and goddesses, especially in their affairs with mortals. Athena, on seeing this, became so furious that she began to beat Arachne until the young girl ran away, placed her neck in a noose, and tried to hang herself. "Live," said Athena, "but hang forever," and Arachne was changed into a spider suspended by a thread. This tale shows the horrible fate that awaits those who would challenge the gods and goddesses. But just a moment! Take a good look at Arachne. Many people think she is creepy, and many others think she is beautiful. Nobody, however, really thinks she appears unhappy.

Was the punishment of Athena so terrible? After all, Arachne not only eluded death but was able to continue the work she loved until the end of time. Animal gods, and tribal totems are far more ancient than divinities in the form of men and women. I suspect that Arachne, far from being a simple mortal, was at one time a divinity who was more powerful than even Athena. The spider symbolizes archaic mother-goddesses, the weavers of fate. This creature is associated with the Egyptian Neith, the Babylonian Ishtar, and the Germanic Holde. In Greek mythology, the three fates, to whom even the greatest of the gods and goddesses are subject in the end, resemble spiders. Perhaps in some lost version of her story, Arachne simply assumed a human appearance to trick Athena, and she revealed her true form after her victory.

The spider is among the most primordial and powerful divinities in many cultures. For numerous West African tribes, the spider is a trickster and a culture hero. Among the Hausa of West Africa, the spider is Gizo. Among the Ashanti of Ghana and in Jamaica, it is Anansi. According to the Ashanti, the animals were once arguing about which was the oldest and deserved the most respect. Finally, they asked Anansi to be the judge, and the deliberations went something like this:

The parrot said, "I was around when there were not yet any blacksmiths. I had to pound iron with my beak, and it is bent to this day."

The guinea fowl said, "I was around when there was a great fire at the beginning of the world. I had to stamp it out with my feet, and they are red to this day."

The elephant, rabbit, and porcupine all told stories from the beginning of the world, showing their great age as well.

Finally, Anansi said, "I was around before earth was created, so there was nothing to stand on. When my father died, there was no ground for a graveyard, so I had to bury him in my head."

The animals bowed to Anansi, acknowledging that he was the most ancient of them all. Boasting aside, the spider—with its thin, segmented limbs—can appear old indeed.

Because spiders are utterly unique, they mirror human alienation from the natural world. The ability to spin threads from their bodies is shared by silk-worms and caterpillars, but spiderwebs are unrivaled in terms of intricacy. It may well be that early hunters were inspired by these webs to create nets and traps. By preserving food to be eaten later, spiders seem to show a human sort of foresight. By stunning prey yet not killing it immediately, they also seem to show a human sort of cruelty.

The spider is a solitary creature, and it has been an aid and inspiration to people who are isolated and pursued. According to legend, when David was fleeing from the soldiers of King Saul and took refuge in a cave, a spider immediately covered up the entrance with its web. The pursuers thought nobody could have entered, and so they passed on. The same story is told of Mohammed when he was hiding from his enemies in Mecca. When Robert the Bruce of Scotland was hiding from the English in a barn on the island of Rathlin, he looked up and saw a spider try six times to swing from one rafter to another. "Now shall this spider teach me what I am to do," said Robert, "for I also have failed six times." The spider made it on the seventh try. Robert returned to Scotland, rallied his men, and won a great victory over the English at Bannockburn in 1314.

But the asocial character of spiders also contributes to the fear they inspire. In the Middle Ages spiders were a frequent ingredient in witches' brew; they were also familiars of witches. The spider, lying in wait for its prey, became a common symbol of the Devil. The most feared spiders of all are the large, hairy ones known as tarantulas, found in Latin America, Africa, and southern Europe. In southern Italy during the Renaissance, there was an epidemic of hysteria about these spiders that was far out of proportion to the danger from their actual bite. People believed that only continual movement could overcome the poison of the spider, and thus, to stay alive, a bitten person had to do a dance known as the "tarantella."

In cultures of East Asia, spiders are generally disliked because they hide in corners. Furthermore, as major predators in the world of small creatures, they seem sinister and almost cannibalistic. Wu Ch'eng-en's *Journey to the West* tells how the monk Tripitaka Tang was once captured by spider-women on his journey from China to India to obtain the Buddhist scriptures. He stopped at a mansion to ask for a vegetarian meal and was greeted by four pleasant young women, but the meal they offered turned out to be human flesh. When the monk tried to leave, they tied him with strings spun from their navels. Only rescue by his animal companions prevented him from becoming their meal.

In Japan there are many stories of enormous spiders that haunt abandoned castles and other ruins. They may take the form of human beings in order to fool the unwary. In one tale collected by Lafcadio Hearn, a samurai went to spend the night in an old temple that villagers said was haunted. A priest came in the night and played on a stringed instrument with more-than-human skill. After a while, the priest turned to the samurai and said laughingly, "Did you think me a goblin? I am only a priest, but I must play my instrument to keep the goblins away. Would you like to try?" The samurai carefully reached out his left hand to touch the instrument, whereupon the strings changed into a giant spiderweb, and the priest became an enormous spider. The samurai drew his sword with his right hand and slashed at the goblin-spider, which retreated. Bound in the web, the samurai could not follow. The next morning villagers came and set him free, and then they followed the trail of blood and killed the spider.

Illustration of a spider in a nineteenth-century book of fairy tales.

By contrast, for many Native American tribes, including the Navaho and Hopi, Spider Woman is the creator deity. According to Navaho legend, a young girl once saw a line of smoke emerging from a hole in the ground. When she looked closely, she saw Spider Woman, who invited her to come down and learn to weave. Navaho women still place a hole in their blankets in memory of the place where the young girl encountered the goddess.

Similarly powerful but far less beneficent is Iktomi, the major trickster figure of the Sioux and other Indians of the American Midwest. According to the traditions of the Lakota Sioux, he is the creator of time and the inventor of language. He is, however, also a coward, a liar, a lecher, and constantly prone to trouble. Undisciplined though he may be, the Indians still respect his power. An offering of tobacco to Iktomi can bring success in hunting.

Like many tricksters, the spider can inspire intense scorn or affection. There could hardly be a more enthusiastic arachnophile than Thomas Muffet, the Englishman who wrote *The Theater of Insects* in 1658. He observed of the spider:

> When she sticks aloft with her feet cast every way, she exactly represents a painted star. As if nature had appointed not only to make it round like the heavens, but with rays like the stars, as if they were alive. The skin is so soft, smooth, polished and neat, that she so precedes the softest skin'd maids, and the daintiest and most beautiful strumpets, and it is so clear that you may almost see your face in her as in a glass; she hath fingers that the most gallant virgins desire to have theirs like to them, long slender, round of exact feeling, that there is no man, nor any creature that can compare with her.

Spiders may be ruthless toward flies, but, Muffet argues, they provide many services to human beings. Not only are their webs good for binding wounds, but the spiders themselves may be used in many kinds of medicines.

Of course, not everyone agreed with his judgment. Perhaps the most famous of the Mother Goose nursery rhymes is a satire that was probably written about Patience Muffet, the daughter of Thomas:

Little Miss Muffet

Sat on a tuffet (a nonsense word),

Eating her curds and whey;

There came a big spider,

Who sat down beside her

And frightened Miss Muffet away.

By giving the spider so many feminine virtues, perhaps Thomas Muffet had made her an ironic patron for girls and women.

Jonathan Swift, however, showed us a distinctly masculine spider in his story "The Battle of the Books," (1697), a satire on the dispute over which writers were better, those of antiquity or those of contemporary times: "For upon the highest corner of a large window there dwelt a certain spider, swollen up to the first magnitude by the destruction of infinite numbers of flies, whose spoils lay scattered before the gates of his palace, like human bones before the cave of some giant. The avenues to his castle were guarded with turnpikes and palisades, all after the modern way of fortification." Because he lived indoors and practiced a sophisticated form of architecture, this spider became the advocate of the modern writers, while a wayward bee represented the ancient ones.

Folklore not only makes spiders fearsome but also gives them protection. It is very widely believed that to kill a spider brings bad luck. Another English nursery rhyme goes:

If you wish to live and thrive,

Let the spider run alive.

A small black spider found in England is known as the "money spider," and if it lands on your clothes, it is an omen of wealth. You must not carelessly brush this spider off, though you are permitted to toss it over your shoulder. The spider is close to primordial powers and should be treated with care.

As chthonic figures, spiders are constantly linked with the dead and the realm beneath the earth. The Massachusetts Puritan Jonathan Edwards preached in his sermon "Sinners in the Hands of an Angry God" (1734) that "God that holds you over the pit of Hell, much as one holds a spider, or some loathsome insect, over a fire, abhors you, and is dreadfully provoked." Spiders are among our most potent symbols of primeval life, yet they are associated not with expansive landscapes but with desolate crannies. This paradox is the basis of a vision by the perverse Svidrigaylov in Fyodor Dostoyevsky's novel *Crime and Punishment* (1866). Svidrigaylov pictures eternity not as vast but as "a little room, something like a village bath-house, grimy, and spiders in every corner . . ."

The spider has a very positive image in the classic children's story *Charlotte's Web*, by E. B. White (1952). By writing messages in her web, Charlotte tricks people into sparing the life of a piglet named Wilbur from slaughter, but her compassion is tempered by acceptance of a natural order in which life is sustained only through killing. Even Wilbur must accept that Charlotte and her children sustain their lives by eating insects.

Giant spiders, sometimes created through radioactivity, have become a cliché of horror and science fiction. A popular comic-book hero today is Spiderman. He climbs buildings and throws out mechanical webs. Although Spiderman is usually a hero, his arachnid identity suggests power, mystery, and a piquant sense of menace. Today, the spider has become a symbol of technology, the gatekeeper of the Internet and the World Wide Web.

3
SAGES

THE PREHISTORIC CAVE PAINTINGS OF FRANCE AND SPAIN ARE AMONG the most ancient works of art that we have. The human beings in these paintings are usually crude stick figures that the artists do not seem to have considered very important. The animals are painted with far more care and passion. The first clearly identifiable religious shrines in history are at Çatal Huyuk in Anatolia and date from around the middle of the seventh millennium BCE. They were dedicated to animals, especially bulls, but also to vultures, foxes, and others. The Egyptians worshipped their gods as incarnated in cats, ibises, bulls, and many other creatures. Over millennia, anthropomorphic goddesses and gods slowly replaced the animal deities. The archaic divinities accompanied their more human successors, often as mascots or alternate forms. Athena, for example, was pictured with an owl, and Zeus with an eagle; Odin was accompanied by ravens and by wolves. The monkey Hanuman, who fought alongside the hero Rama in the epic *Ramayana*, is now perhaps the most popular figure in the Hindu pantheon. In folktales and fairy tales throughout the world, the human hero is often guided or advised by an animal companion. An aura of secret wisdom at times seems to embrace all animals, perhaps a legacy of these bestial deities. Their apparent lack of speech contains more wisdom than our incessant, and often futile, talk. The gaze of many animals, from insects to tigers, often seems more intent and confident than that of any human being. A very few animals, such as the raven or salmon have, however, an especially consistent reputation for wisdom.

The snake, whose reputation for wisdom had already been established when the tale of Adam and Eve was written down, might have been included in this chapter. It has been placed instead in chapter 13 on "Cthonic Animals." The elephant, another candidate for inclusion here, is in Chapter 17 on "Behemoths."

BEE AND WASP

Ask the wild bee what the Druids knew.

—Scottish saying

Our word "bee" ultimately goes back to the Indo-European "bhi," meaning "to quiver." The same root is in the Greek "bios," meaning "life." A quiver is the motion of spirit, a pulse or a breath. Life is a sort of "buzz," a humming in the void. Bees appear primordial. In ancient Egypt, bees sometimes represented the soul. People said that bees were born from the tears of the sun god, Ra, or, later, from those of Christ.

Bees do not fly directly from one place to another. Instead, they often hover pensively in the air. They build complex homes, almost like human cities. Above all else, their ability to produce honey and wax has always seemed wondrous. In Plato's dialogue "Phaedo," Socrates suggested that those who live as good citizens might be reincarnated as bees or other social insects. He meant this transformation, of course, to be a reward. It is small wonder that author Maurice Maeterlink, in the early twentieth century, considered bees the most intelligent of animals next to humankind.

In *Georgics*, Virgil bestowed great praise on the bees, hoping to shame his decadent countrymen in Rome:

> Alone of living things they hold their young
> In common, nor have individual homes:
> They pass their lives beneath the might of law:
> They know the patriot's zeal, and reverence
> For household gods: mindful of frosts to come,
> They toil through summer, garnering their grains
> Into the common store. While some keep watch. . .

A worker bee.

*(Drawn by
Ludwig Becker)*

According to Virgil, bees neither lost their minds through love nor weakened their bodies through sexual excess. They were spared the pains and hazards of pregnancy, since their young spring spontaneously from plants. Most of all, Virgil admired the patriotism of the bees. Spiders, hornets, worms, and other menaces constantly threatened their hives, yet the bees never ceased their vigilance. The individuals sacrificed their lives but the community survived.

Aristotle, writing in *The Generation of Animals*, had considered the generation of bees "a great puzzle," but he suggested various possible means of reproduction. One was that they drew their young from various flowers, while others included copulation and spontaneous generation. Virgil told us that the Egyptians near Canopus by the Nile had a rite in which priests would lead a two-year old bull into a small room. Priests would club the animal to death then continue to pound the flesh, taking care not to pierce the skin. Then they covered the body with thyme, bay and other spices. A short time later, bees emerged from the body.

The practice began after the nymph Eurydice, beloved of Orpheus, departed from her husband into the underworld. The bees died, and they could only be brought back when the spirits of the lovers were placated by the sacrifice of an ox. Persephone, queen of the dead in Greco-Roman mythology, returned yearly from the underworld as vegetation; Eurydice returned as a swarm of bees. Initiates into the mysteries of Dionysus, god of wine and religious ecstasy, had another interpretation of the ceremony; Dionysus, in the form of an ox, had been torn apart by Titans and was reborn as a bee.

Like agriculture, beekeeping is very seasonal. Many bees die and hives lie virtually dormant in winter. In the ancient and Medieval worlds, farmers and their children would watch carefully in spring for signs that bees were beginning to swarm; then people would gather and follow the bees. They would set up attractive new hives and beat on kettles, believing that the noise would help the bees to settle down. In autumn, they would harvest the honey.

Bees are so beloved that people have generally forgiven their painful sting. In a fable of Aesop, however, the bees begged Zeus for stings in order to protect their honey. Zeus was displeased at their covetousness. He granted the request but added that bees had to die whenever they used their stings. Honey bees really do pass away on stinging, since they cannot remove the stinger without tearing their abdomens. (That is not true of other varieties.) The individual dies for the hive; even a person stung by bees might be moved to forgiveness by the sacrifice.

Sometimes armies have set loose bees against their enemies. Montaigne reported in "Apology for Raymond Sebond" that, when the Portuguese were besieging the town of Tamly, the defenders brought a great number of hives and placed them around the wall. Next, they set fires to drive out the bees into the invading host. The enemy was completely routed. On returning, they found that not a single bee had been lost. He does not record how they counted the bees.

The keeping of hives for honey bees developed simultaneously in several centers of the ancient world, including Egypt, Mesopotamia, Greece, and China. It reached a high degree of sophistication by the first millennium CE, but bees, even today, can't be called "domesticated." Even in hives made

The bees here have formed a feudal state, in which the monarch looks on as soldiers and workers parade past.

(J. J. Grandville, from Scènes de la vie privée et publique des animaux, 1842)

by human beings, bees always retain a life of their own. Aelian, writing in the first century CE, reported that bees knew when frost and rain were coming. When bees remained close to their hives, beekeepers warned the farmers to expect harsh weather. If the bees know this, what else might they know as well?

Swarms of bees were closely watched as portents in ancient Rome. Priests used the size and direction of a swarm to foretell fortunes during war. Māra, the Hindu god of love, has a bowstring made of bees, perhaps because discovering love is a bit like a swarm of bees setting off to find a new home. Herodotus, the Greek historian, tells how Onesilus once led a revolt of the Cyprians against the Persian Empire of King Darius. After Onesilus was killed in battle, the people of the Cypriot city of Amathus, who had sided with the Persians, cut off his head and placed it over one of their city gates. After a while, the skull became hollow, and a swarm of bees filled the head with honeycomb. The Amathusians consulted an oracle, who told them to take down the head, bury it, and offer sacrifices every year to Onesilus.

Country people in Europe and North America have traditionally notified the bees when the owner of their property died. There is a description of this ceremony in *Lark Rise to Candleford*, a fictionalized account by Flora Thompson of her childhood in a poor rural family in England during the latter nineteenth century. After her husband's death, his widow, Queenie, would knock on each hive, as though at a door, and say, "Bees, bless, your master's dead, and now you must work for your missis." She told young Flora that, if this were not done, all of the bees would have died. At times, rural people would tell the bees a lot more about affairs of the household, and it is not terribly hard to understand why. Working alone in the fields, one might easily feel an urge to talk. If no other people were around, one might speak to whatever seemed most human – that is, to the bees. The word "bee" is sometimes used to refer to activities where people work and talk, for example a "quilting bee" or a "husking bee." A rumor is sometimes referred to as a "buzz." European peasants have sometimes believed that ancestors return to their property as bees.

People have always considered the honey of bees a divine food. Zeus was raised on the island of Crete, drinking the milk of the fairy goat Amalthea and eating the honey of the bee Melissa. John the Baptist "lived on locusts and wild honey" (Matthew 3:4; Mark 1: 6-7). Pliny the Elder wrote that bees placed honey in the mouth of Plato when he was a child, foretelling his later eloquence. The same was later said of Saint Ambrose, Saint Anthony, and other holy men. Cesaire de Hesterbach reported, in the early thirteenth century, that a peasant once placed the Eucharist in a hive, hoping it would inspire the bees to produce more honey. Later, he found

A queen bee giving out bread and honey to her children in the hive.

(Illustration by J. J. Grandville, from Scènes de la vie privée et publique des animaux, 1842)

that the bees had made a little chapel of wax. It contained an altar on which lay a tiny chalice with the Host.

It is hard for people to think of bees as individuals, and even beekeepers can hardly ever distinguish between them. Only the queen stands out from the rest. Social insects such as ants and bees may be the inspiration for hierarchies in which the individual is subordinate to the state, from ancient Sparta to the Soviet Union. Napoleon took the bee as his emblem. People have constantly aspired to emulate the bees.

Topsell, in the mid-seventeenth century, described the society of bees as an ideal monarchy. The King (now known to be female, and called "the Queen") was set apart by his size and royal bearing. His subjects all loved and obeyed him. The court of bees also contained viceroys, ambassadors, orators, soldiers, pipers, trumpeters, watchmen, scouts, sentinels, and far more.

Thomas Muffet, who collaborated with Topsell has told us that bees "are not misshapen, crook-legged any way, pot-bellied, over close-kneed, bulb-cheeked, great mouthed, lean-chopped, rude foreheads or barren, as many great ladies and noble women are, who have lost the faculty of generation." He went on to say that in the "democratical state" of the bees, everyone is employed in some honest labor. The bees were generated by the putrefaction of animals such as oxen, with the kings and the nobility created from the brain and the commoners from the other parts. Muffet added that bees could not endure the presence of lechers, menstruating women, or those who use perfumes.

For others of the Renaissance, those bees seemed a little too perfect, too virtuous, and too austere. Bernard Mandeville, a Dutch physician, satirized them in *The Fable of the Bees* (1724). The bees appealed to Jupiter to organize their state according to the ideals of perfect virtue. They got rid of the corrupt officials and lazy courtiers. The trouble was that the virtuous replacements didn't know how to get things done. Since the bees no longer produced luxuries like honey, their economy collapsed. Since the bees lived only for peace, they forgot how to fight. Finally, a few melancholy survivors withdrew into a hollow oak to await their end.

The priestesses of the goddesses Demeter and Rhea had been known in ancient Greece as "Melissae" or bees, and the female name hints that, in remote antiquity, people may have realized that the bees are a matriarchy. If so, that knowledge was forgotten until the start of the modern era. It was not until 1637 that the Dutch scientist, Jan Swammerdam, examined the bees under a microscope, to learn that the so-called "King" was really a Queen. The bees could no longer be used as a model for a perfect kingship.

For the French author Maurice Maeterlinck, around the start of the twentieth century, the religion of the bees was that of progress. "The god of the bees is the future," he wrote in *Life of the Bee*. "There is a strange duality in the character of the bee. In the heart of the hive, all help and love each other . . . Wound one of them, and a thousand will sacrifice themselves to avenge the injury. But outside the hive, they no longer recognize each other." The bees had become radical socialists, living only for their cause.

During World War II, the distinguished scientist Karl von Frisch was studying bees at the University of Munich. He had been classified as one quarter Jewish by the Nazi regime, which normally would have deprived him of his job. A disease began to kill off the bees in Germany, threatening the orchards, so the government allowed him to continue his work. Pouring his frustrations into work, the introverted von Frisch began to decipher the communication of the bees. They indicate the direction and distance of food by dances within the hive. This remarkable story, though completely true, reads almost like a fairy tale of grateful animals. The scientist seemed to have a covenant, a special intimacy, with bees. As he worked to rescue bees, they saved him. He was then initiated into their society and their speech. Of course, there is a difference. Unlike the heroes of fairy tales, von Frisch published the secrets of bees to the world.

Still, though a bit of the language of bees may be deciphered, only bees themselves may speak it. A person could certainly try doing the dance of bees, but that would be art. It would not impart information. We teach other animals such as chimpanzees to use human language. Then, much of the time, we think of them as imperfect human beings. But people are also, as Virgil knew long ago, imperfect bees.

For some reason, people pair related animals as opposites: the rat and the mouse, the dog and the wolf, the lion and the tiger. The bee is constantly paired with the wasp, related creatures that live in nests instead of hives. One legend from Poland told that when God created bees, the devil made wasps in a failed attempt at imitation. In a Romanian legend, a peddler persuaded a gypsy to exchange a bee for a wasp by saying that the wasp was larger and would make more honey. All the gypsy got for his greed was a sting.

While the bee is a symbol of peace, the wasp is associated with conflict and war. The Greek comic poet Aristophanes, in his play "Wasps," compared these insects to jurors, since both came in annoying swarms. Saint Paul seemed to conceive of death itself as an insect, probably a wasp, since he asked, "Death, where is thy sting?" (1 Corinthians 15:55). At the end of the Middle Ages, the wasp was frequently the form in which the soul of a witch flew about at night.

A fashionable young lady with a "wasp waist," who is charming but also dangerous

(Illustration by J. J. Grandville, from Les Animaux, 1868)

Some cultures, however, have admired the martial qualities of wasps. Greek warriors went off to battle with wasps emblazoned on their shields. Among Native Americans and Africans, enduring the stings of wasps can be a test in initiation ceremonies. The wasp was a form often taken by Native American shamans. The wasp killing a grasshopper became a Medieval symbol of Christ triumphing over the Devil.

The mason wasp is a deity of the Ila people in Zambia. They tell that once all the Earth was cold, so the animals sent an embassy up to heaven to bring back fire. The vulture, eagle, and crow all died on the journey. Only the wasp arrived to plead, successfully, with God. Because he brought fire to the hearths, the wasp now makes his nest in chimneys.

Today in the United States, the acronym "WASP" stands for "White Anglo-Saxon Protestant." The term is usually used in a derogatory way. It suggests a combination of snobbery, and viciousness. A wasp gives little warning yet has a terrible sting. The term "wasp waist" is used to describe people, especially women, who have what is also called an "hourglass figure." That suggests beauty that is obtained in a very artificial, calculating sort of way. But if the bee were not thought of as holy, perhaps the wasp would not be so maligned.

CROW, RAVEN, AND ROOK

> *One is for sorrow.*
> *Two is for mirth.*
> *Three is a wedding.*
> *Four is a birth.*

> —AMERICAN NURSERY RHYME about crows.

Corvids, particularly ravens and crows, are creatures of paradox. Their black plumage, slouching posture, and love of carrion sometimes make them appear morbid, yet few if any other birds behave in such a playful manner. Even their voices are at once harsh and spirited. Ravens are much larger than crows. They are relatively solitary and make their nests far from human beings, while crows generally move about in flocks and are attracted to human settlements by the promise of food. Both, however, are associated with death and share a reputation as birds of prophecy. They are also monogamous, making them symbols of conjugal fidelity. People probably did not distinguish sharply between ravens, crows, rooks, and many related birds in the ancient world, and they often appear much the same in heraldry as well. The blue jay is one corvid that is not black, but it shares their reputation as a trickster among the Chinook and other Native Americans along the northwest coast of the United States and Canada.

The ambivalent character of ravens is apparent in the Bible where, though "unclean" birds, they sometimes appear to have a special intimacy with God. After the Flood had raged for forty days, Noah sent out a raven to find land. It flew back and forth until the waters receded but did not return (Genesis 8: 6-8). Later, however, ravens fed the prophet Elijah every morning and evening when he had fled from Ahab into the wilderness (1 Kings 17:4). In the Talmud, when Abel had been slain, Adam and Eve, who had no experience with death, did not know what do. A raven slew another of its own kind, dug a hole, and performed a burial, thus demonstrating to the first man and woman how the dead ought to be treated. In gratitude, God feeds the children of the ravens, which are born white, until they grow black plumage and can be recognized by their parents.

There is a similar story in the Koran. When Cain had killed Abel, he did not know what to do with his brother's body, and carried it around on his shoulders, for no human being had ever died before. Then Cain saw a raven scratch the ground, showing how people should bury their dead. On giving Abel a proper burial, he was able to repent and attain divine forgiveness.

The Romans viewed birds as mediators between gods and human beings, at times in homey as well as solemn ways. Pliny the Elder recounts a story of a raven, hatched on the roof of a temple in Rome, which flew down to the shop of a shoemaker. The owner, wishing to please the gods, welcomed the bird. By watching the customers, the raven soon learned to talk. Every day he flew to the podium across from the forum and greeted the Emperor Tiberius by name. Then he flew around and said "hello" to various men and women before returning to the shop. One day a neighbor killed the raven, perhaps thinking the bird had left some droppings on his shoes. The people of Rome became incensed and lynched the man. Then they gave the raven a splendid funeral, in which Ethiopian slaves carried the bier, and many people left flowers along the path.

On the European continent, where there are few vultures to compete with, corvids would hover above a battlefield and later descend to eat the corpses. Two ravens perched on the shoulders of the Norse Odin, who was intimately associated with battles. They were named Hugin (thought) and

Munin (memory), and they flew over the world to bring news to the god. The Celtic war goddess Morrigan would take the form of a raven or crow and come as a herald of death. When the hero Cuchulainn had been mortally wounded, he tied himself to a tree and stood with his sword in hand. His enemies watched from a distance but did not dare approach until a crow, the goddess Babd, perched on his shoulder. In one of many versions collected from oral traditions, the traditional British ballad "The Twa Corbies" begins:

> There were three ravens on a tre,
> They were as black as black might be:
> The one of them said to his mate,
> "Where shall we our breakfast take?"—
> "Downie in yonder green field,
> There lies a knight slain under his shield . . .

The ravens find they must take their meal elsewhere, for this knight is guarded by his dogs, hawks, and wife. In many wars, however, it gave soldiers a sense of foreboding to see corvids following their armies and hovering over the battlefield.

Some major themes such as prophecy and death run through the lore of ravens among Europeans, Asians, and Native Americans, but the tales of the Indians are richest in humor. Among the Haida Indians and related tribes of the American and Canadian Northwest Coast, Raven is at once a sage and trickster. One widely disseminated tale, which has many variants, begins at a time when there was no light in the world, for all of the light was held in a box kept in the house of the Chief of Heaven. Raven conceived a plan to steal the light. First, Raven turned into a cedar leaf floating in a stream where the daughter of the Chief of Heaven went to drink. On swallowing the leaf, she became pregnant, and then gave birth to Raven, in the form of an infant boy. After playing for several days in the house of the Chief, Raven began to cry and clamor for the box that held the light. The Chief, who was charmed by his young grandson, let him hold the box, at which point Raven immediately put on wings and carried the container out of the Chief's lodge into the sky. An eagle pursued him, and Raven swerved to escape its talons, dropping half of the light, which broke into many fragments, becoming the moon and stars. Finally, when he had reached the edge of the world, Raven let go of the remaining light to create the sun.

It is possible that this story may have been influenced by Christianity, since it has many structural parallels with the story of Jesus. The Chief of Heaven corresponds to God the father, while his daughter is reminiscent of Mary, and Raven himself may be Christ, referred to in the Bible as "the light of the world" (John: 8:12). But in such tales, the identity of Raven—with his cosmic powers and many changes of form—becomes so abstract that he seems less an animal, or even a person, than an embodiment of celestial energy. The myths of Raven often resemble accounts of contemporary physicists, who describe how the cosmos was created, in narratives full of mysterious forces and alternate universes.

Corvids have always figured prominently in poetry, and one of the most famous examples is Edgar Allan Poe's poem "The Raven" (1845). The narrator asked a raven that had flown into his chamber whether he could be reunited with his deceased beloved:

> "Prophet!" said I, "thing of evil! prophet still, if bird or devil!—
> Whether Tempest sent, or whether tempest tossed thee here ashore,
> Desolate yet all undaunted, on this desert land enchanted—
> On this home by Horror haunted—tell me truly, I implore—
> Is there— is there balm in Gilead? — tell me— tell me, I implore!"
> Quoth the Raven "Nevermore."

The bird gazed imposingly, as befitted a messenger from the world of spirits, but revealed nothing.

In "Vincente the Raven" (1941) Portuguese author Miguel Torga tells a story of the raven that accompanied Noah. Vincente became increasingly restless. Though not personally mistreated, he grew angry that the animals and the earth should be punished for the crimes of humankind. At last he left the Ark unbidden, perched on the peak of Mount Ararat, and called out his defiance to God. The flood continued to rise, but Vincente refused to leave. God realized that, should he drown Vincente, his creation would no longer be complete, and so he finally relented and reluctantly allowed the water to recede.

In the collection of Medieval Welsh tales known as *The Mabinogion*, the giant Bran (whose name means "Crow"), traditionally depicted with a raven, had been mortally wounded while leading an army of Britons against the Irish. At his command, his followers behead him and carry the head to the site of the Tower of London for burial, so that it might serve as a charm to protect Britain. This is the remote origin of the legend that Britain will never be successfully invaded as long as ravens remain in the Tower. Such pagan legends eventually led to the demonization of crows and ravens at the end of the Middle Ages, when they were often either witches' familiars or a form in which witches flew about at night.

Contemporary culture is full of urban legends of vanishing hitchhikers and bigfoots, but these generally do not address momentous themes such as the destiny of a people, and so we hesitate to call them "myths." One story that does, perhaps, merit such a designation is that of the ravens in the Tower of London. At least six ravens are kept on the grounds of the Tower of London at all times, because of a reputedly "ancient" prophecy that "Britain will fall" if they leave. In fact, the ravens were brought to the Tower of London in 1883, to serve as props for tales of Gothic horror told by Beefeaters to tourists. A Yeoman Warder might intone, " . . . and, after they chopped off her head, ravens descended to pluck out her eyes," and a real raven would be croaking in the background. The legend that ravens protect Britain from catastrophe actually dates only from 1944, when they were used as unofficial spotters for enemy bombs and planes. But, since the story still addresses visceral anxieties, it continues to circulate and develop. Since it encourages respect for ravens, as avatars of the natural world, we should be glad.

Rooks share the reputation of the more illustrious ravens for wisdom, but they are more approachable. In *Precious Bane* by Mary Webb (1924), a novel of peasant life in the English countryside during the early nineteenth century, a family told the rooks when the master of the house had died, so that the birds would not bring ill luck by deserting the home. The new master

*The raven tells the fool "cras,"
Latin for "tomorrow," leading
him to procrastinate.*

*(Albrecht Dürer's illustration
for Sebastion Brandt's
Ship of Fools, 1494)*

followed the ceremony cynically, remarking quietly that he was very fond of "rooky pie." The birds rose and circled thoughtfully but then returned to their branch, letting the people know they intended to stay. Their hesitation, however, left a sense of foreboding, and the farm was soon struck by disaster.

The crow even teaches people how to die in a myth of the Murinbata, an aboriginal people of Australia. Crab demonstrated what she believed was the best way to die by going to a hole and casting off her wrinkled shell. Then she waited for a new one, so she might be reborn. Crow responded that there was a quicker, more efficient way to die, and immediately fell over. The lesson here was that the reality of death should be accepted without equivocation.

Herodotus wrote that two "black doves" had flown from Thebes in Egypt; one settles in Lybia while the other goes on to Greece and settles in the sacred grove of Dodona. It rustles the leaves and brings forth the prophetic voice of Zeus. The Greek historian believed the birds were originally dark-skinned priestesses, but scholars have suggested that they may have been crows or ravens.

Closely bound with the wisdom of corvids is their reputation for longevity, and they can indeed live for decades. In *The Birds*, by the Greek comic playwright Aristophanes, crows are said to live five times the life of a human being. In a dialogue entitled "On the Use of Reason by so-called 'Irrational' Animals" by Plutarch, the wise pig Gryllus states that crows, upon losing a mate, will remain faithful for the remainder of their lives, seven times that of a human being. Precisely because of this reputation for fidelity, however, the Greeks and Romans considered a single crow at a wedding to be an omen of the possible death of one partner.

The god Apollo once took the form of a crow when he fled to Egypt to escape the serpent Typhon. The crow remained sacred to Apollo, but the relationship of the god to corvids was not without ambivalence. Ovid tells in *Fasti* that once Phoebus Apollo had been preparing a solemn feast for Jupiter and told a raven to bring some water from a stream. The raven flew off with a golden bowl but was distracted by the sight of a fig tree. Finding the fruits unfit to eat, the raven sat beneath the tree and waited for them to ripen. He then returned with a snake that he claimed had blocked the water, but the god saw through this lie. As punishment for lateness and deceit, the god later decreed that the raven would not be able to drink from any spring until figs had ripened on their trees. The god placed a constellation depicting a raven, snake, and bowl in the sky, and, in the springtime, the voice of the raven is still harsh from thirst. The call of the raven was often said to be "cras," Latin for "tomorrow," and the raven often symbolized a procrastinator during the Renaissance.

The intelligence of crows and ravens has amazed people since ancient times. A fable about their smart feats is In "The Crow and the Pitcher," traditionally attributed to the legendary Aesop, a thirsty crow came upon a pitcher of water but was unable to reach inside and drink. The bird began to pick up pebbles and drop them one by one into the pitcher until the water had risen to the top. The usual moral of this story is that "Necessity is the mother of invention." This is one anecdote that could well be based on fairly accurate observations. That particular feat was long considered

Aesop's fable
"The Crow and the Pitcher."

(Illustration by Richard Heighway)

impossible for birds, but in 2009 several rooks in a Cambridge, England aviary figured out how to make the water level rise by dropping pebbles in a test tube.

In China, crows can be birds of ill omen but also symbols of fidelity in love. A collection of Taoist lore usually entitled *Strange Stories from a Chinese Studio*, written in the latter seventeenth century by Po Songling, tells of a young man from Hunan named Yü Jung who failed his civil service examinations and was consequently unable to find employment. Desperate and hungry, Yü Jung stopped at the shrine of Wu Wang, the guardian of crows, and prayed. After a while, the attendant of the temple approached and offered him a position in the order of the black robes. Delighted to have found a way to earn his living, Yü Jung accepted. The attendant gave him a black garment, and, after putting it on, Yü Jung was transformed into a crow. Soon he married a young crow named Chu Ch'ing, who taught him the corvid ways. Unfortunately, he proved too impetuous, and a mariner shot him. The others crows churned up the waters and made the mariner's boat capsize, but Yü Jung suddenly found himself once again in human form, lying near death on the temple floor. At first he thought the whole adventure had been a dream, but he could not forget the joys he had known as a crow. Eventually, he passed his exams and became prosperous, but Yü Jung continued to visit the temple of Wu Wang and make offerings to the crows. Finally, when he sacrificed a sheep, Chu Ch'ing came to him and returned his black robe, and Yü Jung again took on corvid form.

In the story "Herd Boy and the Weaving Maiden," popular in many versions throughout East Asia, corvids come to the aid of lovers. The daughter of the King of Heaven, who wove the silk of clouds, had married a humble herdsman, and the two spent so much time together that they neglected their duties. The King finally placed the Weaving Maiden in the western sky and the herd boy in the eastern sky, where they were separated by the river of the Milky Way. One day every year, the crows and magpies gather and form a bridge across the sky so that the lovers may be briefly reunited.

In the Ghost Dance religion founded by the Paiute Indian shaman Wodvoka near the end of the nineteenth century, the crow was the messenger between the world of human beings and that of spirits. Indians from many tribes in the southwest of the United States, together with some whites, engaged in an ecstatic dance to bring about the regeneration of the earth. The celebrants wore crow feathers, painted crows upon their clothes, and sang to the crow as they danced. Sometimes they sang of Wodvoka himself flying about the world in the form of a crow and proclaiming his message.

In a volume of poetry entitled *Crow*, British poet Ted Hughes constructed a personal mythology. A figure named Crow continually does battle with cosmic powers; he may be defeated or victorious but always survives. But ravens and crows are not at all endangered. Corvids are found nearly everywhere in the Northern Hemisphere from remote cliffs and forests to cities. They neither fear man nor need him, and their resilience constantly inspires our respect.

OWL

Now the wasted brands do glow,
Whilst the screech-owl, screeching loud,
Puts the wretch, that lies in woe,
In remembrance of a shroud.

—WILLIAM SHAKESPEARE, *A Midsummer Night's Dream*

Anybody who has come across the eyes of an owl shining alone in the night knows why they have always been associated with the dead. This is true especially in Northern regions where many owls are at least partially white; their feathers eerily reflect the moonlight. Not only does the owl look like an apparition, but its call is usually a drawn-out, wavering note that easily suggests the muffled voice of a spirit. Owls are attracted to places of burial by the smell of decaying flesh, and they could easily have been taken for the spirits of the deceased. These birds also have exceptionally fine sight at night. In total darkness, they can navigate by hearing or, with some species, even by echolocation. The ability of owls in flight to locate mice far below them, even under a coating of snow, still sometimes impresses researchers as almost supernatural. The ancient Egyptians represented part of the soul, called the "ba," as a bird with a human head, and pictures of it in *The Egyptian book of the Dead* somewhat resemble an owl. In mythology and literature, death is intimately associated with wisdom, and the owl is an ancient symbol of both.

With most varieties of owls, the female is a bit larger than the male, which may partially explain why owls often symbolize primeval feminine power. In the Sumerian poem "The Huluppu Tree," the goddess Lilith made her home in the hollow of a tree, but the hero Gilgamesh cut the tree down to make a throne for Inanna, the Queen of Heaven. Owls often nest in such places, and Lilith, who became a demon in Hebrew tradition, is referred to as a "screech owl" in the book of Isaiah. Babylonian reliefs from the early second millennium showed a goddess-demon, probably Lilith, with the claws and wings of an owl, sometimes with owls at her side. Ever since, owls

Sept, a Chtonic God of Egypt associated with the Owl.

have been the frequent companions of sorceresses and goddesses. Athena, the Greek goddess of wisdom and war, is also closely associated with the owl. Homer refers to her as "owl eyed," and the earliest depictions represent her as a woman with the head of an owl. She was later depicted holding an owl aloft in one hand. The Latin word for "owl," "strix," is the origin of "striga" or "witch." In *The Golden Ass* by Lucius Apuleius, written in the first century CE Rome, a witch flies off at night in the form of an owl.

The owl is a loner among the birds, a figure of both awe and revulsion. In one tale from the Hindu-Persian *Panchatantra*, the birds were so impressed by the owl's venerable demeanor that they elected him king. During the day, when the owl was asleep, the crow mocked the choice, saying that the owl was repulsive with his hooked nose and huge eyes. The birds rescinded their decision, but an enmity between owl and crow remains to this day. This tale probably alludes to the way crows and other birds mob predators such as owls and usually succeed in driving them away.

In the Middle Ages the owl often represented the Jews, for, like them, the bird was said to have "scorned the light." Writers of Antiquity including

Pliny the Elder and Aelian had observed that other birds mobbed the owl when it appeared during the day. This idea was used later in Christian Europe as a justification for attacks on Jews who ventured beyond their ghetto.

For Odo of Chiteron, a clergyman writing in Kent during the early thirteenth century, the owl represented the rich and powerful who abuse their position. In his fable "The Rose and the Birds," he recounts how the birds came upon a rose and decided that it should go to the most beautiful bird. They then debated whether this should be the dove, the parrot, or the peacock, but had come to no decision when they went to sleep. The owl stole the rose during the night, and so the other birds banished him from their presence; they still attack him if he shows himself by day. "And what will happen on judgment day?" Odo continued. "Doubtless all the angels . . . and just souls will—with screams and tortures—set upon such an owl." It is a conclusion that anticipates the peasant's revolts and other revolutionary movements that would start to become more common towards the end of the Middle Ages.

The aristocrats, who could be contemptuous of the masses, occasionally took the solitary nature of the owl as evidence of its superiority to other birds. A late 15th century shield from Hungary, now in the Metropolitan Museum of Art in New York, shows an owl, perched above the coat of arms of a noble family, saying, "Though I am hated by all the birds, I nevertheless enjoy that."

At the same time, the image of the various birds ganging up on the owl very clearly suggested the persecution of Christ. The full ambivalence that people felt towards the owl was expressed in an anonymous Middle English poem from the early twelfth century, entitled *The Owl and the Nightingale*, a heated debate on many subjects between the nocturnal bird of prey and the beloved songbird. The nightingale accused the owl of filthy habits, and the latter replied that she cleaned the churches and other buildings of mice. When the nightingale taunted the owl by saying that men used his dead body as a scarecrow, the owl replied that he was proud to be of service after death. (The image of the stuffed owl, however, suggests a crucifix, which was also a sort of scarecrow set up to keep demons away.) The owl boasted of being able to foresee the future and warn people of impending disaster. No judgment was rendered in the poem, but most readers think the owl got the better of his adversary. Medieval artists sometimes placed a cross above the head of an owl to indicate that it represented the Savior.

The idea that the cry of an owl foretells death is found in a remarkable range of cultures, from the Greeks and Romans to the Cherokee Indians. For the Navaho, the owl was a form taken by ghosts. For the Kiowa, it was a favored form of magicians after death. The Pueblo Indians would not enter a house where owl feathers or the body of an owl was displayed. For the Aztecs and Mayans, owls were the messengers of the god of death, Mictlantechtli. In the Aztec rites of human sacrifice, the heart of the victim was placed in a stone container decorated with an owl. Among several West African tribes such as the Yoruba, the owl is often a form taken by evil magicians, and simply to see or hear and owl can bring ill luck. For the Chinese, the great horned owl was the most powerful symbol of death.

In the modern period, however, life expectancy expanded dramatically, and people were no longer so constantly reminded of their mortality. In consequence, literature began to emphasize the reputation of the owl for wisdom rather than for death. Since the nineteenth century the "wise old owl" is probably most familiar as a figure in books for children. In the Disney film *Bambi*, an owl is even shown benevolently instructing baby rabbits and other creatures of the forest, which it would normally devour.

In J. K. Rowling's Harry Potter stories, wizards communicate by sending letters by owl. Harry, a budding young wizard who is learning magic at Hogwarts School, has an owl named Hedwig who keeps him company during lonely evenings. When not working as his messenger, she flies freely in and out of Harry's room, sometimes bringing back dead mice, and affectionately nibbles on his ear.

Nevertheless, the fear that owls traditionally arouse has by no means vanished entirely. Occasionally in the late twentieth century, a proposal has been raised to control the population of rats in New York City by importing owls. This has never gotten very far, in part because the call and eyes of an owl during dark urban nights have proved too unsettling to people.

CARP AND SALMON

> *Now I am swimmer who dies,*
> *Who runs with rain and moon and salt-wind tide,*
> *River and falls and sweet pebble water.*

Kwakiutl Indian tale, "Swimmer the Salmon" (adapted by Gerald Hausman)

Since very ancient times, people have thought of the sea as the womb from which all life emerges. Fish have symbolized the inexhaustible fertility of nature, and they have been closely associated with mother-goddesses such as Tiamat and Atargatis. Fishermen are the last hunter-gathers, and, even in antiquity, they were already surrounded by nostalgia and romance. The first disciples of Jesus were fishers, and Christ told them, "Follow me and I will make you fishers of men" (Matthew 4:19). The fish was the earliest symbol of Christ, and also the first avatar of the Hindu deity Vishnu. But despite their enormous importance in religion and other aspects of human culture, fish are remarkably difficult to humanize. The reason may be a combination of their remote, expressionless eyes and their utter silence, which contrast with the constant speech and expressive glances of human beings. Even in the animistic world of folklore, a talking fish is only rarely found, and when a fish does speak, it is usually in connection with some remarkable event.

The salmon, however, lives according to a remarkably "human" pattern. It is born in fresh water, migrates to the ocean and finally returns, sometimes swimming hundreds of miles upstream, to the place of its birth to spawn. Atlantic salmon may sometimes make the journey a few times during their lives, but Pacific salmon die after laying their eggs. In their determination to complete their destiny, salmon seem not only human but also very noble indeed. The entire life of a salmon may be understood as a sort of quest. It follows the mythic pattern described by Joseph Campbell in *The Hero with a Thousand Faces*, where the archetypal hero, after many adventures, returns to end his life in the place of his birth. The way the salmon crosses the boundary between fresh water and the sea suggests a passage between the realm of men and immortals, and thus the salmon is a symbol of transcendence in many cultures.

Fish carved on reindeer antlers have been excavated in Paleolithic settlements of Spain and France, and a few of these fish are clearly recognizable as salmon. According to the Norse Eddas, Loki, the god of fire, assumed the form of a salmon in order to hide beneath a waterfall after he had offended the other deities with his taunts. But the salmon is most central in the cultures of the Celts and the Indians of North America.

The salmon of wisdom, which has superhuman knowledge, appears often in the myths and legends of the Celts. In the tale of "The Marriage of Culhwch and Olwen," this is the salmon of Llyn Llyw, one of the oldest animals in the world. When King Arthur and his knights sought the hunter Mabon, they consulted this salmon, which not only told them where Mabon was imprisoned, but also ferried two men there upon its back.

According to one Irish legend, the old poet Finneagas once caught the salmon of wisdom and told the hero Finn MacCumhail to roast it for him. A blister arose on the salmon, and Finn pressed it, burning his thumb, which he placed in his mouth to ease the pain. Then Finn was suddenly filled with wisdom, which was apparent in his gestures and his gait. Finneagas had been expecting to gain this wisdom himself by eating the salmon, but he realized that destiny had granted it instead to his pupil, and so he was happy to have mentored Finn.

Several tales place five salmon of wisdom in Connla's well, near Tipperary. Above the well were nine hazel trees, and their purple nuts would fall into the well to feed the salmon. The hazel nuts represented the spirit of poetry, and the sound of them striking the water was said to be lovelier than any human song. The bellies of the salmon were purple from the nuts, and their wisdom constantly increased. Only the salmon might eat the nuts safely, yet legend tells that the goddess Sinend was once so eager for wisdom that she defied the prohibition and approached the well too closely; the waters then rose and swallowed her.

A similar tale is told of a young girl named Liban in *The Book of the Dun Cow*, from the early Middle Ages. A well overflowed to form a lake, and

Liban was swept to the bottom with her dog, but God protected them from the waters. The two stayed there for a year, until Liban saw a salmon and prayed, "O my Lord, I wish I were a salmon, so that I might swim with the others through the clear green sea!" In that moment, she became a salmon from the waist down, while her dog became an otter, and together they swam about for three hundred years. Finally, she allowed herself to be captured by some holy men and be taken to a cloister. She died immediately after being baptized and was consecrated as one of the holy virgins.

The return of a salmon to the place of its birth has suggested the coming of Christ, and the transition between Celtic religion and Christianity was probably eased by the shared symbolism of the fish. A similar significance was eventually accorded in Celtic areas to eels, which share with salmon the ability to move between pond and ocean. The salmon of knowledge could also be the ultimate origin of the wise fish in such popular fairy tales as Grimms' "The Fisherman and his Wife" where it is a flounder, and Alexander Afanas'ev's "Emilya and the Pike."

The salmon also represents rebirth among Native American tribes of the Northwest Coast such as the Kawkiutl and Haida, not, however, as a unique event but as part of an eternal cycle. The salmon swimming upstream as it endeavors to elude predators such as bears signifies the individual bravely endeavoring to complete his or her destiny. Finally, after spawning, the dead salmon are swept back into the ocean, symbols of the ultimate union with all of life.

The carp is essentially a fresh-water fish, but it also swims upstream, sometimes leaping over falls in order to spawn, and Asian deities are often depicted riding upon carp. According to Chinese legend, the carp, on reaching its final destination, will become a dragon. In East Asia, the carp is a symbol of perseverance, and it is used especially to signify the scholar who studies hard to pass his examinations. The bright scales of a carp resemble armor, so the carp was often used as a symbol of samurai warriors. The Japanese have a national holiday in early spring known as "Children's Day," in which they fly carp-shaped kites, in order to inspire boys and girls to show persistence and courage.

To an extent, the carp shares the salmon's reputation for mysterious wisdom. The first recorded versions of the enormously popular fairy tale "Cinderella" came in the early ninth century in China, and the helper of the young girl—the equivalent of the fairy godmother in the well-known version by Charles Perrault—is a fish. The young girl takes the fish home from a well, to be kept in a pond. Her wicked stepmother kills the fish out of spite, but the heroine prays to its bones, which grant her every wish. The kind of fish was not specified, but, since it was brightly colored and kept as a pet, it certainly appears to have been a carp.

Carp were introduced to Europe from East Asia in about the fifteenth century, and they have greatly extended their range throughout the world with the growth of trade. Salmon, by contrast, are now endangered everywhere, in part due to excessive fishing but mostly as a result of dam construction along their migratory routes. Varieties of salmon are produced for the market in hatcheries, artificially bred and even genetically engineered. These are sometimes inadvertently released into the wild, where they interbreed with wild salmon, often further endangering the original inhabitants of streams. In contact with human beings, it may be safer for animals to be beautiful than useful or wise.

Ostriches

4
JUST BEAUTIFUL

WE GENERALLY PLACE "BEAUTY" IN THE PROVINCE OF ART RATHER than that of nature. In his celebrated *History of Beauty*, Umberto Eco has nothing at all to say about the beauty of animals, plants, or landscapes by themselves. They become relevant for him only when portrayed in paintings, photographs, or poems. We can legitimately call many, perhaps all, animals and plants "beautiful," but, for that very reason the adjective does not exalt any much above the rest. A few mammals such as the giraffe and hind are especially known for their grace. The creatures that are particularly famous for their beauty, however, are mostly birds. A few such as the peacock and swan are so overlaid with human symbolism that they belong as much to fairy tales as to fields and lakes, as much to culture as to nature.

One animal that could have been included in this chapter but has been placed elsewhere is the dove, found in Chapter 18, "Divinities."

OSTRICH, PARROT, PEACOCK, AND HUMMINGBIRD

> *Remember that the most beautiful things in the world are the most useless, peacocks and lilies for instance.*
>
> —JOHN RUSKIN, *The Stones of Venice*

Today, it is primarily women who wear the brilliant plumage of birds, whether on hats or in jewelry, though that has not always been the case.

Among many indigenous peoples such as those of the Americas and New Guinea, male warriors have traditionally worn the most opulent feathers. The association of women with plumage does, however, go back to ancient times, specifically to Egypt. Maat, the goddess of cosmic harmony, was one of the few Egyptian deities that were generally depicted in fully human form, and she wears a large ostrich feather in her hair. Though styles of dress and appearance are notoriously fickle, she appears elegant, dignified, and remarkably contemporary about three millennia after the paintings were executed. When souls were judged in the next world, the heart of the deceased would be weighed against Maat, sometimes represented by an ostrich feather, to decide whether the person would join the immortals or be devoured by demons.

Nevertheless, the fashionable plumes often received more respect than the bird itself, which was often described in very bizarre ways. Ancient writers such as Pliny and Aelian maintained the ostrich was able to eat and digest virtually anything, even stones. Pliny added that the bird buried its head in the sand to evade danger, an erroneous idea that has by no means disappeared today. According to Medieval bestiaries, the ostrich could divine when to lay eggs by watching the stars. !Kung Bushmen accord the eggs of an ostrich supernatural power, since they are so hard that no animal's jaws are able to break them. In Europe, these were often taken for the eggs of a griffin and placed in royal treasuries. Sometimes the "griffin eggs" were made into goblets, in the belief that they would change color when touched by poison.

The hummingbird has been admired not only for its bright, iridescent colors but also for its speed. It is able to hover in one place, switch directions instantly, and fly backwards, giving it the sort of maneuverability valued by military commanders. Despite, or partly because of, its diminutive size, this bird was an attribute of the Aztec god of war, Huitzilopochti, whose headdress was made of hummingbird feathers. The Mayan deity Quetzalcoatl, in his incarnation as a serpent, also had the plumes of a hummingbird. Battle seems to be almost universally associated with eroticism, and plumes of the hummingbird have also been frequently used in love charms.

The parrot of myth and legend is generally a peaceful bird, but one that seems almost too splendid for this world. In the Hindu-Persian *Pan-*

The first man and woman are living peacefully among the animals before the Fall, where the cat does not seem to notice the rat in front of it. Note especially the parrot on a sprig in the hand of Adam, probably a branch from the Tree of Life, which he will carry when exiled from Paradise.

(Adam and Eve, by Albrecht Dürer, 1504. Metropolitan Museum of Art, Fletcher Fund, #MM2218B)

chatantra, the god Inidra kept a parrot of extraordinary beauty and intelligence named Blossom. One day, as the parrot was sitting in the palm of his master's hand, Yama, lord of the underworld, appeared. Blossom was terrified, and the deities begged Yama to spare the parrot. The dark figure replied that the decision was not his to make, so the gods took their petition to Death himself. On beholding the visage of the Grim Reaper, the parrot perished in terror.

The ancient Egyptians kept rose-ringed parakeets in cages as pets, but the real popularity of parrots in the Mediterranean area dates from the time

*The parrot, with its
brilliant plumage, has
often been associated
with royalty.*

(Illustration by J. J. Grandville)

of Alexander the Great, who brought back previously unknown varieties
from India. They attained new popularity with the vast expansion of maritime trade at the end of the Middle Ages. Much of this exploration was
motivated by a longing to find a place of primal innocence, perhaps even
the original Eden. In an engraving of 1504 entitled "Adam and Eve," the
German artist Albrecht Dürer showed a parrot perched on a sprig from the
tree of life, held by Adam as he speaks with Eve in Paradise.

For people in the Northern Hemisphere, the visual splendor of parrots
has always seemed to be heightened by their exoticism and remarkable ability
to mimic the human voice. According to Medieval legend, a parrot had
announced the coming of the Virgin Mary, and artists often painted exotic
birds beside her and the infant Jesus. One Cardinal during the Renaissance
reportedly paid a hundred gold pieces for a parrot that could clearly recite
the entire Apostles' Creed. On the other hand, some people have always under-

stood innocence as an invitation to corruption; rough mariners enjoyed teaching parrots obscenities. Commerce in parrots was from the start a rather shady business, in which theft and smuggling were common, and it now threatens to drive many of the remaining species of parrots to extinction.

Perhaps the most opulent bird of all, the peacock, is originally from India but was imported to the Mediterranean world in very early times. The spreading tail of the peacock is an ancient symbol of the sun, its feathers standing for rays of light. The peacock was sacred to Zeus and, most significantly, to his wife Hera. According to one myth, after changing the maiden Io into a heifer as punishment for her affair with Zeus, Hera sent the hundred-eyed giant Argos to watch over her. Taking the form of a woodpecker, Zeus signaled the location of Argos to Hermes, who then killed the monster. Hera took the eyes of Argos and, as a memorial, placed them in the tail of a peacock. The story is a bit ironic, since the peacock spreads his tail as part of a mating dance. The eyes, then, are a warning to young ladies not to trust the macho posturing of men.

The peacock, however, became ever more a symbol of splendor over the centuries. It was the mount of Karrtikeya, the Hindu god of war, as well as that of Brahma and his wife Sarasvati. The kings of Persia sat on a peacock throne, and Chinese Emperors from the Ming dynasty onwards bestowed peacock feathers as a sign of favor. Aristocratic gardens throughout Eurasia, which were meant to be a recreation of Paradise, had peacocks strolling about the grass. In Christianity, the peacock became a symbol of the resurrection, as well as of the all-seeing eye of the Church. In the modern period, however, as governments became democratic and people increasingly distrusted royalty, the peacock gradually became a symbol of vanity. Finally, as advertisers revived the abandoned heraldic symbols of an earlier era, the peacock was used to represent the wonders of color in film and television.

Pileated woodpecker. The woodpecker is not a singer so much as a musician, but its sound announces the start of the rainy season in many cultures.

5
MUSICIANS

MUSIC MAY BE A VERY INTIMATE EXPRESSION OF WHAT IT MEANS TO BE "human," yet it is also something we share with many other animals, especially with birds and insects. While there is still some debate about to what extent the call of a thrush should be considered "music," the designation goes far beyond being just a metaphor. The academic field of biomusicology centers on comparisons between human music and the sounds of animals. Both have similar structures, patterns, and rhythms. But these, in turn, have analogs elsewhere in the natural world such as the patterns in sedimentary rocks or the rings of trees trunks, which show a comparable blend of repetition and variation. Composers have transcribed patterns of DNA with musical notes, and found that it sounds rather like music of the Baroque period. There seems, in summary, to be something universal in music, an appeal that transcends the boundaries of species. The taste in music by other creatures varies, however, with culture. The Greeks, for example, preferred the sound of cicadas and other insects, while the British of the nineteenth century especially loved songbirds.

The music of whales became especially popular in the late twentieth century, but they have not been included in this chapter. Instead, they may be found in Chapter 17, "Behemoths and Leviathans," which seemed even more appropriate.

CICADA, GRASSHOPPER, AND CRICKET

The poetry of earth is ceasing never:
On a lone winter evening, when the frost
Has wrought a silence, from the stove there shrills
The cricket's song, in warmth unceasing ever,
And seems to one in drousiness half lost,
The grasshopper's among some grassy hills.

—JOHN KEATS, "On the Grasshopper and the Cricket"

Entomologists place grasshoppers, locusts, mantises, and crickets in a single order, the o*rthoptera*, while cicadas belong to the order *homoptera*. But modern taxonomies do not necessarily reflect popular perceptions of animals today, much less the ways in which creatures have been regarded over centuries. For the ancient Greeks, grasshoppers, locusts, crickets, and (sometimes) mantises went under the single name of "akris," and modern translators of their works have to determine which insect seems most appropriate from the context. All of these insects were often difficult to see and were known to people primarily through their sounds in open fields. These noises, created by rubbing parts of their bodies together, are often amazingly loud for the tiny creatures that produced them, and they are often synchronized as well. They are mating calls, produced almost exclusively by males, and some ancient myths suggest a surprising awareness of this fact. In the case of *orthoptera*, people also knew them through the enormous damage they often did to crops, only partially compensated for by the favor these insects themselves enjoyed in culinary traditions.

The sound of the grasshopper is not particularly melodic, but it is very steady. The grasshopper provides a regular beat to accompany the singing of birds and other creatures of the field. Even today, when we have almost entirely ceased to mark dates by the behavior of wildlife, the incantation of the grasshopper still marks the beginning of autumn for many people, and the cessation of it signals the coming of winter, at which time we begin to see the bodies of perished grasshoppers along country roads. One of the most famous fables is "The Ant and the Grasshopper," colorfully retold by the British folklorist Joseph Jacobs in *The Fables of Aesop*:

A concert of insects at a wedding party.

*(J. J. Grandville, from
Scènes de la vie privée et publique des animaux,
1842)*

In a field one summer's day, a Grasshopper was hopping about, chirping and singing to its heart's content. An Ant passed by, bearing along with great toil an ear of corn he was taking to the nest.

"Why not come and chat with me," said the Grasshopper, "instead of toiling and moiling in that way?"

"I am helping to lay up food for the winter," said the Ant, "and recommend you do the same."

"Why bother about winter?" said the Grasshopper; "we have got plenty of food at present." But the Ant went on its way and continued its toil. When the winter came the Grasshopper had no food, and found itself dying of hunger, while it saw the ants distributing every day corn and grain from the stores they had collected in the Summer. Then the Grasshopper knew . . .

It is best to prepare for the days of necessity.

The contrast between the two insects is essentially that between an artist and a laborer, settled this time to the latter's advantage.

The locust is the Mr. Hyde to the grasshopper's Dr. Jekyll. Desert locusts usually look and behave much like grasshoppers, but in conditions of crowding and scarcity of food, they undergo a metamorphosis, changing their body color and growing larger. They also begin to reproduce more quickly and to swarm, laying waste to the surrounding countryside with incredible speed. The Bible tells us that when the Pharaoh refused to let the people of Israel leave Egypt, locusts were the eighth plague sent by Yahweh in punishment:

The locusts invaded the whole land of Egypt. On the whole territory of Egypt they fell, in numbers so great that such swarms had never been seen before, nor would be again. They covered the surface of the soil till the ground was black with them. They devoured all the green stuff in the land and all the fruit of the trees

(Exodus 10:14-15).

In North Africa and the Near East, plagues of locusts have continued to occur into the twentieth-first century, sometimes darkening the sky and confirming the general accuracy of the Biblical descriptions. In similarly vivid terms, the prophet Joel compared locusts in their vast numbers and their destructiveness to an invading army. The same imagery was used by the Egyptians themselves, and in an inscription commemorating the deeds of Ramesses II—according to some traditions the very Pharaoh who was confronted by Moses—at the battle of Kadesh; it reports that the armies of the Hittites covered the mountains like locusts. For all the trouble they caused, however, the Egyptians do not seem to have hated or despised such insects, and one text from the Old Kingdom speaks of a ruler ascending to heaven in the form of a grasshopper.

In an Islamic folktale from Algeria, the Devil looked scornfully on the newly created world and said, "I can do better than God." "Very well," replied God, "I will give you the power to bring to life whatever creature you create. Stroll about the world and return in a hundred years." The Devil took up the challenge and put together a creature with the head of a horse, the breast of a lion, the horns of an antelope, the neck of a steer, and parts from several other animals. Since the parts did not fit properly, he began to whittle away at the creature, until only a tiny locust was left. The Lord says, "Oh, Satan . . . What is this! To show your impotence and my power I will send swarms of this creature around the earth, and thus I will teach people that there is only one God."

Perhaps because locusts in the Bible were always a scourge of God, the insects themselves have usually not been heavily stigmatized. People will hardly ever eat creatures that they find repugnant, except in times of severe hunger, but locusts and grasshoppers are eaten in much of Africa. They are also mentioned as a possible food in the Bible (Leviticus 11:20-23).

In ancient Egypt, the grasshopper was a popular motif, often depicted on festive items cosmetic boxes or jewelry. The prophet Isaiah said of Yahweh, "He lives above the circle of the earth; its inhabitants look like grasshoppers" (Isaiah 11:24). These insects have continued to be used as symbols of insignificance, in ways that may be either endearing or contemptuous. It is possible to see a bit of each response in the Greek myth of Tithonus, a prince of Troy, who was loved by Eos, goddess of dawn. The deity asked Zeus to grant her lover immortality, but she forgot to ask for eternal youth for him as well. As he grew old, she left him, and eventually he withered away and became a grasshopper.

The repeated sounds of crickets have not always impressed Westerners as unequivocally beautiful or cheerful. In Germany, somebody with a neurotic obsession is said to have crickets in his head. On the other hand, repetitions can represent the sometimes irritating yet essential lessons of conscience. One variety is known as the "house cricket" for its habit of frequently entering homes. Because these crickets are drawn to warmth, they are symbols of the hearth. To have such a visitor is traditionally considered good luck throughout Europe, and killing it can bring ill fortune. The Chinese, however, value crickets for their martial spirit, and gladiatorial combats between crickets have been a popular sport in China since ancient times.

In Carlo Collodi's classic for children *Pinocchio* (1883), the hero, a wooden puppet that has come to life, smashes a cricket named Jiminy with a mallet but, after many misfortunes, regrets his evil deed. Disney Studios later made Jiminy Cricket into one of their most popular animated characters, and even had him introduce the television show *Walt Disney Presents*.

Cicadas, especially, are generally known only through their sound, since they dwell in trees and are usually not seen until they die and fall to the ground. For the Greeks, they seemed to be incorporeal beings and symbolized immortality. In Plato's dialogue "Phaedrus," Socrates and the young man Phaedrus had been engaged in a passionate discussion of philosophy, and the former remarked that the cicadas, while singing and conversing among each other, must surely also be observing their dialogue. He went on to explain that these insects were once human beings. When the Muses first came to earth, bringing with them music and the other arts, a few people were so thrilled with their new gifts that they would do nothing except sing, quite forgetting to eat and drink. After a while, they died happily, thinking only of their art. They returned to life as cicadas, creatures to which the Muses have granted the boon of never needing any earthly sustenance. They sing from their moment of birth onwards, without food or drink, until the day of their death, after which they go and report about human beings to the Muses, telling who has honored the arts and their divine patrons. Socrates assured his young pupil that the two of them might expect a good report from the cicadas, since they had been discoursing on the theme of love.

These insects have traditionally had a similar meaning among the Chinese, who also believed that cicadas lived only on dew. From the late Chou through the Han dynasties (ca. 200 BCE through ca. 220 CE), the Chinese would place jade cicadas in the mouths of their dead to insure immortality. For cultures of the Far East, the songs of insects such as cicadas and crickets also represent the chanting of Buddhist priests. They are sometimes kept in cages, and their songs are often considered more beautiful than those of birds.

CUCKOO, LARK, NIGHTINGALE, AND WOODPECKER

> Now more than ever seems it rich to die,
> To cease upon a midnight with no pain,
> While thou are pouring forth thy soul abroad
> In such an ecstasy!
> Still wouldst thou sing, and I have ears in vain—
> To thy high requiem become a sod.
>
> —JOHN KEATS, "Ode to a Nightingale"

Before the modern era, the sounds of nature were everywhere, day and night. Buildings, even Medieval castles with walls thick enough to resist sieges, were not built to keep them out. The sounds of birds, most especially, were used to mark both the hours of the day and the seasons. The cuckoo is the bird of spring, while the lark sings at early morning, and the nightingale during the night. This gave them significance at once practical and poetic, as is illustrated by this exchange from Shakespeare's *Romeo and Juliet*, taking place after a night of love:

> JULIET:
> Wilt thou be gone? it is not yet near day;
> It was the nightingale, and not the lark,
> That pierced the fearful hollow of thine ear;

While the nightingale proclaims his love, the rose cavorts with a dirty beetle.

(Illustration by J. J. Grandville, from Les Animaux, 1868)

Nightly she sings on yon pomegranate tree:
Believe me, love, it was the nightingale.

ROMEO:
It was the lark, herald of the morn,
                    No nightingale.

These two birds were constantly used to signal the time of day and night. The association of birdsong with hours is why many of the first affordable clocks used a mechanical cuckoo to announce the time.

The song of the cuckoo traditionally announces the beginning of summer with an outpouring of exuberant energy. Farmers understood it as a signal to begin planting, but spring is above all the season of love. Through most of history, apart from the High Middle Ages and the nineteenth century, amorous passion has been regarded with suspicion, and that may also be said of the cuckoo. Its song has been traditionally a good omen to those who planned to marry, but also a warning of possible adultery to those already wedded.

Pliny the Elder suggested that hawks transformed themselves into cuckoos, since the hawks seemed to vanish at about the same time cuckoos became numerous. He observed, however, that hawks would eat cuckoos if they did meet. The idea reflected the cuckoo's reputation for treachery, since, as Pliny put it, "the cuckoo is the only one of all the birds that is killed by its own kind." This superstition continued into the twentieth century in parts of Europe.

According to one myth, Zeus first made love to Hera after he had moved her to pity by appearing in the form of a disheveled little cuckoo. The bird was an attribute of Hera and adorned her scepter. Indian poets knew the cuckoo as the "ravisher of the heart," and the god Indra also assumed the form of a cuckoo for the purpose of seduction.

The idea that the cuckoo is an adulterer has at least some distorted basis in observation, since the European cuckoo will lay its eggs in the nest of another bird. The egg containing the young cuckoo will generally hatch first, and the fledgling will push the other eggs from the nest. Pliny explained this by saying that all other birds so hated the cuckoo that they dared not make nests, for then they would be vulnerable to attack. The only way the cuckoo can procreate is by concealing the identity of its offspring.

The use of the word "cuckold"—derived from the Old French "*cocu*," meaning "cuckoo"—for a man whose wife is unfaithful goes back at least to the late Middle Ages. Shakespeare wrote in his play *Love's Labor's Lost*:

> When daisies pied, and violets blue,
> And lady-smocks all silver hue,
> Do paint the meadows with delight,
> The cuckoo then on every tree
> Mocks married men; for thus sings he,
> "Cuckoo;
> Cuckoo, cuckoo" O word of fear,
> Unpleasing to the married ear!

In an era when marriage for love was still a somewhat revolutionary idea, the cuckoo increasingly came to represent sexual desire, while the nightingale signified romantic love.

While the cuckoo of literature is masculine, the nightingale is usually female in Western culture, and people have found her song less exuberant than sweet or sad. Her tragedy, as told by Apollodorus, began when Procne, a princess of Athens, married King Tereus of Thrace, and they had a son named Itys. Tereus raped Philomela, Procne's sister, and then cut out the tongue of the victim so she would not reveal his crime. Philomela wove a tapestry depicting the deed, made the cloth into a robe, and gave it to Procne, who then killed Itys, boiled him, and served his son up to Tereus in revenge. When the King realized what had happened, he set out in pursuit of the two sisters. The women prayed to the gods, who then turned Procne into a nightingale, Philomela into a swallow, and Tereus into a hoopoe. Later Latin authors, however, confused the two sisters and called the nightingale "Philomela," a name afterwards used by poets throughout Europe, perhaps because the song of a nightingale seemed less that of a killer than that of an innocent victim. According to Pliny the Elder, their song was so beloved in Rome that caged nightingales in Rome commanded the sort of prices usually paid for slaves.

In traditions of the Near East, the nightingale is masculine and in love with the rose, a tragic passion incapable of consummation, but the Islamic world shared Western ambivalence about romantic passions. In *The Conference of Birds*, written by Sufi poet Farid Ud-Din Attar in Persia around the end of the twelfth century, the hoopoe summoned the birds to a pilgrimage to their king, the simorgh. The nightingale answered that roses flowered only for him, and so he could not leave even for a single day. The hoopoe replied that the love of the rose was a superficial illusion, and that the rose really mocked the nightingale by fading in a day.

In the tenth-century Arabic fable *The Island of Animals*, however, the nightingale proved to be the most eloquent and sensible of the animals. He surpassed even such fine speakers as the jackal and the bee, as the beasts brought suit for mistreatment against people before the King of the Jinni. When a man from Mecca and Modena argued that human beings were especially favored by God, the nightingale carried the day by replying that human beings therefore had a special responsibility not to abuse other creatures.

In Russia, by contrast, nightingales were often associated with witchcraft. There was a great demand for caged nightingales to sing in the homes of aristocrats and wealthy merchants. Peasants hired to capture the birds would have to wander about the woods at night following the sound, and they often feared becoming victims of enchantment. In Russian folklore, "Nightingale" was a monstrous bird-man who was half bird, nested in oak trees, lay in wait for travelers on the road to Kiev, and could whistle up a wind strong enough to kill human beings.

The lark begins to sing at early morning before the sun has even risen, and so it has been associated with beginnings. In "The Birds" by the Greek comic playwright Aristophanes, the lark boasts that it is not only older than the gods but the very earth itself, an idea perhaps inspired by the ability of the lark to sing in flight. When its father died, there was no ground, so the lark had to bury him in its own head.

As with so many other things, people tend not to appreciate animals until they begin to disappear. As Europe industrialized and many birds became less common, romantic poets of the nineteenth century celebrated their songs with perhaps unprecedented intensity. The singing of birds represented a sort of poetic inspiration that was utterly natural and spontaneous. Among the most famous lyrics of the period were "Ode to a Nightingale" by Keats and "To a Skylark" by Shelley, in which the poets long to enter the world of joy that could inspire the songs of a bird. Hans Christian Andersen celebrated the beauty of nature over the creations of humankind in "The Emperor's Nightingale," which tells of a mechanical bird that fails to sing as sweetly as one in the wild.

An elegiac poem, which perhaps marks the end of this tradition, is "the Darkling Thrush" by Thomas Hardy, published in the first years of the twentieth century, which ends:

> At once a voice arose among
> The bleak twigs overhead
> In a full-hearted evensong
> Of joy illimited;
> An aged thrush, frail, gaunt, and small,
> In blast-beruffled plume
> Had chosen thus to fling his soul
> Upon the glowing gloom.

So little cause for carolings
Of such ecstatic sound
Was written on terrestrial things
Afar or nigh around,
That I could think there trembled through
His happy good-night air
Some blessed hope whereof he knew
And I was unaware.

As the twentieth century progressed, writers increasingly understood references to nightingales or larks as part of outmoded poetic diction.

The woodpecker is not a singer so much as a musician, but its sound announces the start of the rainy season in many cultures. The sound of a woodpecker knocking its beak against a tree is like a martial drum and resonates loudly through the forest. The woodpecker was sacred to Ares, Greek god of war. Romulus and Remus, the legendary twins who founded Rome, were suckled by a wolf and fed by a woodpecker. Ovid wrote in *Metamorphoses* that the witch Circe changed a young man named Picus, son of the Roman god Saturn, into a woodpecker after he had refused her advances.

As more of a fighter than a lover, the woodpecker has never been terribly popular, yet it may do better than songbirds in fitting the raucous popular culture of the latter twentieth century. One of the most popular cartoon characters of the mid-twentieth century was the violent, and frequently amoral, trickster Woody Woodpecker.

In recent decades, the ivory-billed woodpecker, which was once common in the forests and swamps of the Southeastern United States, has attained an iconic status. Due primarily to loss of habitat, it probably became extinct in the latter twentieth century, yet there are still occasional reports of possible sightings, and some birders have made finding it into almost a mystic quest.

6
TOOTH AND CLAW

IN MEDIEVAL AND RENAISSANCE EUROPEAN PAINTING, THE ENTRANCE to the Underworld is often portrayed as an enormous mouth with fangs, sometimes recognizable as that of a highly stylized lion or serpent. This iconography reflects a recognition that, in the natural world, to be eaten is the virtually inevitable fate of all creatures, from the mightiest to the smallest. It is also a testimony to the enormous hold large predators continue to have over human imagination, even for people living in towns or highly cultivated landscapes. Mircea Eliade has traced much of religious experience to "the mystical solidarity of predator and prey" and Christ, for example, is traditionally represented as both a lamb and a lion. The Eucharist, which is the central mystery of Christianity, places believers in the role of predators, and the Deity in (essentially) that of prey. In a more secular context, predators are generally associated with royalty and nobility, and carnivorous animals are prominent in heraldry. As parts of the world grew more democratic, resentment against the mighty was taken out on large predators, which were often hunted to near extinction in the nineteenth century. People have traditionally seen the burying of the dead as a defining moment in the creation of "civilization," through which we attempt to exempt ourselves from the cycle of birth and death that governs the lives of other creatures. According to Judeo-Christian traditions, animals ate only vegetation until after the Flood and the New Covenant. Our profound ambivalence about predators, in summary, impacts our entire relationship with the natural world.

Other animals that might have been included here are the crocodile and the eagle. The former is in Chapter 17, "Behemoths and Leviathans"; the latter in Chapter 18, "Divinities."

LION, PANTHER, JAGUAR, AND TIGER

For he is of the tribe of Tiger.
For the cherub cat is a term of the Angel Tiger.

—CHRISTOPHER SMART, "My Cat Jeoffry"

Lions are social animals that live in prides, in which the females do most of the hunting. They inhabit open plains, though their once vast range is now reduced to the savannas of Africa. Lions are often followed by scavengers from vultures to hyenas, which has contributed to the idea that lions are kings attended by a court. The male has an enormous head and luxuriant mane, which suggests the sun sending forth rays. Tigers, by contrast, are usually solitary, and they are found in the jungles of Asia and the forbidding hillsides of Siberia. Though normally shy near human settlements, they will occasionally attack human beings. Panthers, which are almost identical to leopards apart from the color of their fur, are smaller than either lions or tigers, and they rely more on stealth and speed in hunting. They are able to climb trees, where they can hide meat from scavengers, observe while unnoticed, and pounce suddenly upon their prey. Panthers and leopards are solitary, nocturnal hunters, often associated with chthonic realms.

Even in Paleolithic times, the great cats seem to have had a special religious significance, and they were given a place of honor among the cave paintings of Lascaux in a cavern known as the "Chamber of Felines." At the dawn of urban civilization, people already thought of these animals as primarily feminine. Our words "female" and "feline" both ultimately come from the Latin "felare," meaning "to suck." Several figurines of women, possibly goddesses, accompanied by great cats have been found at Çatal Huyuk in Turkey, the earliest known walled town.

In early pantheons, the great cats are most closely associated with fem-

Illustration by Julius Schnorr Carolsfeld, mid-nineteenth century, to the story of Daniel who was thrown to the lions but remained unharmed.

inine deities. Among the foremost of these was the Egyptian Hathor, who was the goddess of love, dance, and feminine arts but was also capable of great fury. When men rebelled against the sun god Ra, she attacked them as a lioness and soon developed an insatiable thirst for blood. When Ra himself was satisfied that the rebellion had been defeated, she continued to kill, and the gods feared that she would destroy all humankind. They left out vats of red wine, and she drank them, mistaking the liquor for blood, fell asleep, and finally awakened with her anger appeased. Hathor in her incarnation as a furious avenger was known as Sekmet and was depicted with the body of a woman and the head of a lioness. The Babylonian goddess Ishtar, in her capacity as a deity of war, was represented standing upon a lion. Lions were harnessed to the chariot of Cybele, the Syrian goddess who was adopted by the Romans as their Magna Mater.

Male lions, however, are just as common in the visual arts of the ancient world. Both the Egyptians and Mesopotamians placed stone lions as guardians on each side of the doorways to temples and palaces, a practice that eventually spread eastwards all the way to China. In Sumero-Babylonian animal proverbs, which are among the very earliest literary works to have survived, the lion is already established as the king of beasts. This motif soon became one of the most widely established literary conventions, found in fables attributed to the semi-historical Aesop. The lion often appears as a figure of brute power that terrorizes other animals, and the sly fox in one fable observes that many tracks lead into his cave but none lead out. The lion is not always dominant, however, and in another fable the ass and other animals he once tormented beat the aged lion. In African legends as well, the majestic lion frequently falls victim to weaker but cleverer creatures such as the hare. The motif of a lion as monarch has been used in the Hindu-Persian *Panchatantra*, the Medieval European stories of Renard the Fox, the Narnia stories by C. S. Lewis, and countless other works throughout the world.

To kill a lion was a supreme achievement for a warrior in the ancient world, and kings of Egypt and Mesopotamia frequently had themselves depicted hunting lions. Many heroes of the ancient world including Gilgamesh, Hercules, and Sampson, were conventionally depicted wearing the skin of a lion. The Greco-Roman fable attributed to Babrius told of how a lion and man were once travelling together, when they passed a sculpture of a hero strangling the king of beasts. The man pointed to it as proof of human superiority, to which the lion replied that if lions had done the carving, "You would see men victims of lions." Even the Hebrews, who generally despised predators, could not help feeling some admiration for these animals. The lion became a symbol of the biblical Judah, later of Saint Mark the Evangelist, and of Christ himself.

Romans imported vast numbers of lions for their gladiatorial games, where these creatures represented the Emperor as they devoured criminals before a raucous audience. Since lions do not readily attack human beings, the animals were starved or specially trained for the job. According to one popular story, a runaway slave named Androcles who had been recaptured was placed in the area with a lion. Instead of devouring him, the lion licked his feet. A leopard was immediately let loose, but the lion killed it. The

Emperor Drusus ordered the slave to come forward and asked why the lion had spared him. Androcles told how, after extreme mistreatment, he had escaped his master and taken refuge in a cave. The lion had come to Androcles and held up a bloodstained paw, from which the fugitive pulled out a thorn. From that time on, the lion had fed him, bringing Androcles part of every kill. Moved by the story, Drusus granted both Androcles and the lion their freedom. (The circuses relied on theatrics almost as much as on blood, and this incident could possibly have been staged, in order to make the Emperor seem magnanimous.)

The Romans did have remarkable skill at training lions. According to Pliny, Mark Anthony harnessed lions to his chariot and rode around with a courtesan, an event that shocked his contemporaries and may have contributed to his eventual downfall. The Emperor Caracalla later had a pet lion that sat by his table, slept in his bedroom, and was even kissed by him in public. In the Christian era, Saint Jerome, who lived as a hermit in a cave, reputedly tamed a lion in the same way as Androcles, and thus a lion was traditionally painted at his feet.

In heraldry, the lion has always represented royalty, and it is often depicted wearing a crown. This symbolism was even adopted in areas that had been free of lions since Paleolithic times, such as China and Western Europe. Precisely because actual lions were unknown outside of a few royal menageries, it was easy to stylize these animals into symbols. King Richard I, known as "Lionheart" for his courage in battle, chose three golden felines against a red background for the coat of arms of England. Though some scholars believe these were originally intended to represent leopards, people universally regard them as lions today. The English lion and the Scottish unicorn now flank the heraldic emblem of Great Britain.

Eventually, the lion became so deeply associated with the institution of kingship that it was almost impossible for most Europeans to think of the animal in any other context. Monarchists liked to imagine the lion as dignified to the point of blandness, and they excused the predatory nature of the beast by saying he would only kill as much as he needed to eat. In the modern era, however, democratically inclined people often stigmatized lions as vicious.

The lion king surrounded by obsequious courtiers.

(Illustration by J. J. Grandville, from Fables de la Fontaine, 1839)

The nineteenth-century French romantic Eugene Delacroix painted bloody battles between lions and Arabs, celebrating the ferocity of both. In the same era, the lion tamer became a feature of large circuses. He would be a burly man with a handlebar moustache, often wearing leopard-skin trunks, who would compel lions or other big cats to obey by cracking a whip. The spectacle dramatized the ability of humanity to control nature, and, by analogy, the dominance of men over women.

The tiger was a more unequivocally romantic beast, admired almost as one might admire a storm or volcano. Malaysian myths tell of a city that tigers built entirely from human skin, bones, hair, and other body parts. The Hindu Kali, goddess of time, who wears a necklace of skulls and holds a sword of destruction in one of her many hands, has often been portrayed riding upright upon a tiger. In his aspect as a destroyer, the god Siva wears the skin of a tiger. We should remember, however, that the annihilating power of these figures was viewed not as evil but simply as part of the cosmic cycle.

In China, the tiger, ruler of the earth, has often been paired with the dragon, ruler of the sky, as the two greatest primordial powers. The dragon creates clouds, while the breath of the tiger becomes wind, and together they bring rain. The tiger is associated with autumn, since it resembles that season in its violence and destruction; its black and orange coat also reminds people of fallen leaves. The tiger, rather than the lion, is called "king of beasts" in Asia. It is the third sign of the Chinese zodiac and is often depicted with wings. Patriarchs of Taoism have been represented riding upon tigers, signifying their ability to live in harmony with the elements.

The tiger entered Western imagination as Alexander the Great invaded India. The god Dionysus, sometimes identified with Alexander, has been occasionally depicted in a chariot pulled by tigers when crossing the Tigris River on his way to India. The Romans, who were as much attracted to exoticism as violence, had tigers in their circuses.

Pliny the Elder reported how people accomplished the seemingly impossible task of capturing tiger cubs. The captor would steal several cubs, jump on a fast horse, and ride until the tigress would start to overtake him. Then he would drop one cub, forcing the tigress to pause and take it back to her lair. She would then start off again after the captor, and this might be repeated several times until at last the man would reach his ship with a single cub. This story was frequently repeated in Medieval bestiaries, but with an additional twist. The horseman steals only one cub, but he throws glass balls to the tigress, and she mistakes her reflection for a cub.

Medieval depictions of Eden usually included the lion but almost never the tiger. The tiger was too unequivocally frightening to be used much in heraldry, but it again entered Western awareness when India became part of the British Empire. Thomas Bewick wrote towards the end of the eighteenth century that the tiger "fears neither the sight nor the opposition of man . . . and it is even said to prefer human flesh to that of any other animal." For the British colonists and many of their Indian supporters, extermination of tigers became a humanitarian mission, and many boasted of having killed hundreds. Meanwhile, the most intense opposition to British rule in Southern India came from Tipu Sultan who believed that it was "better to live two days as a tiger than two hundred years as a sheep." He sat on a throne decorated with tigers, had the stripes of a tiger placed on the uniforms of his soldiers, and emulated a tiger's reputed cruelty.

It was at this time that Blake wrote "The Tyger," perhaps the most famous animal poem of the modern era, which began:

> Tyger! Tyger! Burning bright
> In the forests of the night,
> What immortal hand or eye
> Could frame thy fearful symmetry?
>
> In what distant deeps or skies
> Burnt the fire of thine eyes?
> On what wings dare he aspire?
> What the hand dare seize the fire?

Tales of the cruelty of Tipu Sultan, which were filtering back to England, probably influenced Blake, and he might have seen the tigers on display in the menagerie of the Tower of London. The illustration that Blake painted to accompany the poem, however, shows far more affection than awe. It is clearly a domestic housecat, though perhaps he intended to show how every pussycat has a tiger inside.

The beauty and terror that are inextricably blended in the tiger are expressed in a parable called "The Lady and the Tiger," which anonymously passes through our culture like a legend—everybody has heard it, yet hardly anybody knows where it is from. It was actually written by Frank Stockton and published in *The Century Magazine* in 1882. A king had decreed that justice for a serious crime was to be decided by a test in which the accused was placed in a large arena and, before spectators, choose between two doors. Behind one door was a tiger that would rip him apart with its claws. Behind another was a lady, carefully chosen as a match, whom he would have to marry. The King heard that his daughter and a common man were in love, and he ordered the young man sent to the area to be judged by providence. The princess found out what lay behind the doors, and made a secret signal to her lover. The story ended with the famous question: "Which came out of the opened door—the lady or the tiger?" This sort of choice obsessed Victorians: bourgeois domesticity or unadulterated passion? But the beautiful lady, as Freud recognized, is really an aspect of the tiger, and marriage to her is a sort of annihilation.

In the twentieth century, the tiger has taken over many symbolic values previously associated with the lion. In his poem "Geronition," T. S. Eliot used the tiger as a symbol of Christ. Advertisers have constantly exploited the primitive energy associated with the tiger. Esso Petroleum, for example, has since the mid-1960s advertised gasoline with the slogan "Put a tiger in your tank!" Symbolic importance in human culture, however, often makes animals more vulnerable in the wild, since it means that people will be more drawn to hunt them for folk remedies or sport. The Caspian tiger became extinct in the 1970s and the Javan tiger a decade later, while the few surviving species continue to hold on very precariously.

In the legends of Africans, who had direct experience of both, the lion may often have been the ruler of animals, but the leopard generally inspired greater awe. The black color of the panther enables it to blend into forests, while the spots of the leopard suggest innumerable eyes. Members of powerful, secret societies of Central Africa claim the power to transform themselves into leopards, though they are very secretive about how this is done. The legendary ancestor of the kings of Dahomey and other African lands is a leopard that came out of a river to lie with a woman, and the ferocity of the beast explained their penchant for war. The Biblical prophet Jeremiah, frustrated by the inability of the Hebrews to put aside their wicked ways, asked, "Can . . . the leopard change his spots?" (Jeremiah 13:23). A leopard's skin is a garment traditionally worn by African monarchs. The style has been taken up in the west, where leopard spots, suggesting fierceness and status, are often printed on women's garments and accessories.

The leopard or panther was sacred to Osiris, god of the dead, in ancient Egypt. The Greeks identified Osiris with Dionysus, and the priests of both wore the skins of a panther. Panthers generally drew the chariot of Dionysus, and they were sometimes depicted in his entourage. Both of these deities, in turn, often came to be identified with Jesus, and writers of the Middle Ages often praised the panther. Medieval bestiaries told how other animals would follow the panther drawn by the sweetness of its breath. Only the dragon would flee and take refuge in a cave, much as the Devil would run in fear from Christ.

With the end of the Middle Ages, however, many people strove to cleanse Christianity of pagan elements, and the panther, so important in pre-Christian religions, was consequently condemned for viciousness. At the beginning of Dante's *Divine Comedy*, the narrator encounters three predators, a panther, a she-wolf and a lion in a dark wood. The panther is the first to threaten him, but he is saved from the beasts by the intercession of the poet Virgil.

In the early twentieth century, the poet Rainer Maria Rilke lamented the loss of primeval wildness in his poem "The Panther." It describes a panther pacing ritualistically in the zoo, where it can see little besides bars, and concludes:

> Only at times, the curtain of the pupils
> Lifts, quietly——. An image enters in,
> Rushes down through the tensed, arrested muscles,
> Plunges into the heart and is gone.

The iconographic use of the panther in recent times has veered between anger and nostalgia. A militant African-American group that supported armed revolution against the American government in the 1960s and 70s was known as the "Black Panthers."

Perhaps the most mysterious big cat of all may be the jaguar, native to Latin America. The motif of the jaguar appears so often in the arts of early Native South American communities that historians of religion believe it may have been the master of animals, perhaps even the supreme god. In tribes of Bolivia, at least until very recent times, killing a jaguar with a wooden spear has been a test of manhood, used in the initiation of a warrior. Shamans are sometimes believed to be able to turn themselves into jaguars. During an eclipse, people howl to scare the jaguar that is trying to devour the sun. In recent times, however, machines have increasingly taken over the symbolism of animals, and for most people today the "jaguar" is a luxury car.

WOLF

Homo Homini Lupes est. (Man is a wolf to man.)

—PLAUTUS

Perhaps more than any other animal, the wolf has been associated with martial qualities. It has been continually condemned for rapaciousness and cruelty, yet it has also been praised for fierceness. Language contains many traces of lupine totems among several Eurasian tribes in archaic times. Many German names include the root "wolf": "Wolf," "Wolfgang," "Wolfram," "Wolfhart," and others. The common French name "Luc" and the English "Luke" are related to "loup," the French word for "wolf." Names containing "wolf" are also common among the Cheyenne and other Indian tribes of North America, one remarkable testimony to the surprising universality of animal symbolism. The Native Americans admired the wolf not only for its prowess in hunting but also its loyalty to the pack.

A totemic identification with the wolf is perhaps most memorably recorded in the story of Romulus and Remus, the founders of Rome who were suckled by a wolf in their infancy. Initially, the wolf was probably not merely their nurse, but also their mother. Every year on the 13th through the 15th of February, Romans celebrated the Lupercalia, an archaic festival in her honor, on the Palatine Hill by the cave where the infants had been sheltered.

It is also likely that Romulus and Remus were, at one point, werewolves—human beings who transformed themselves into lupine form. Legends of werewolves were common among the warrior clans of the ancient world. *The Iliad* mentions a warrior named Dolan who went about in the guise of a wolf until finally recognized and killed by the Greeks. Herodotus reported a belief that a Nomadic tribe known as the Nueri would change themselves into wolves for a few days every year. Sigmund of the Volsung clan, a hero of Norse and German mythology, would put on a pelt at night to become a wolf. Pliny the Elder reports a belief among the Greeks that the Arcadians would select a man from a certain noble family every year and lead him to a marsh. The man would then strip, hang his clothes on a tree, swim to a desolate area and become a wolf for a period of nine years. The legend probably derived from an initiation ritual into a clan of the wolf. The wolf was also sacred to the god Apollo.

As humanity became increasingly settled and urbanized, the reputation of the wolf declined. Ovid writes in his *Metamorphoses* that Jupiter, the supreme god, once came to King Lykaon in the guise of a simple traveler. As was his practice, the King served up human flesh for his guest. Enraged, Jupiter took on his true form, rose to the sky and hurled a thunderbolt at the palace. The King fled in terror and, as he ran, was changed into a wolf.

Even Herodotus and Pliny, who were often credulous, did not fully believe reports of people changing into wolves. By the time of Christ, the educated people of Greece and Rome generally dismissed stories of werewolves. Such tales appealed, however, to a growing taste for horror. Perhaps the most famous werewolf tale of all is in the *Satyricon* of Petronius, written in the middle of the first century CE. The freedman Niceros accompanied a young soldier to a farm at night. They stopped for a while by a cemetery, where the companion slipped away. After waiting nervously for a while, Niceros looked around and caught sight of the soldier taking off his clothes. The soldier then urinated in a circle, and, after completing this ritual, turned into a wolf, howled, and ran away. Niceros hurried to the farmhouse, where he learned that a wolf had broken into the pasture and attacked the sheep, but one of the farmhands stabbed the beast in the neck with a spear. On returning home, Niceros found the soldier bleeding profusely from the neck, and he knew that his former companion was a werewolf.

Norse mythology shows a highly ambivalent attitude towards the wolf. Two wolves accompany Odin, the god of magic, but he is destined to fall prey to the great wolf Fenris when gods battle giants at the end of the world.

A wolf being arrested for the murder of sheep

(Illustration by J. J. Grandvale from Les Animaux, *1868)*

This monster was sired by Loki, the god of fire, and he grew so strong that the gods themselves were terrified. No ordinary fetter could hold Fenris, so the gods summoned the dwarves to fashion a chain made from the footfalls of a cat, the roots of a mountain, and other mysterious ingredients. Though the fetter appears soft as a silken string, Fenris refused to let himself be bound with it, unless a god placed one hand in its mouth as a pledge. Only Tyr, god of battles, had the courage to do this. After failing to break the chain, Fenris bit off Tyr's hand. As the final days of the world approach, Fenris is destined to escape.

The ancient Hebrews were primarily a tribe of herders, and wolves were a constant threat to their sheep. In the Bible, wolves were often identified with either invading armies or with Hebrews who have become greedy and corrupt. Reproaching his nation, the prophet Ezekiel said, "Her leaders in the city are like wolves tearing their prey, shedding blood . . . " (Ezekiel 21:27). Jesus told his disciples, "I am sending you out like sheep among the wolves . . . " (Matthew 10:16).

St. Francis tamed a wolf, Gubbio, who later accompanied him, a feat Christians often interpreted as the triumph of spirituality over appetite. During the Middle Ages, bounties were placed on wolves, sometimes in the same amounts as those for the heads of brigands and highwaymen. Wolves were hunted to extinction in England during the fifteenth century, in Scotland during the sixteenth, and in Ireland during the eighteenth. The last wolves in Germany were shot around the middle of the nineteenth century. Wolves have remained continuously in mountainous regions of France, Spain, and Italy as well as in the forests of Eastern Europe. With the growing fear of witchcraft in the Renaissance, people began to regard wolves not simply as pests but as agents of the Devil.

In the *Malleus Mallificarum* or "Witch's Hammer," a manual for witch finders (1484), author Heinrich Kramer, and possible co-author James Sprenger, stated that wolves that show "such astuteness that no skill and strength can capture them" must be either devils in disguise or scourges of God. Wolves were regarded as a form in which witches went about at night. Particularly in France, suspected werewolves were often brought to trial and executed. Certain physical features such as eyebrows that grew together were considered signs of a secret identity as a werewolf, and any unexplained wound or scar might arouse suspicion of clandestine adventures in lupine form.

The wolf's reputation for rapacity extended across Eurasia, and stories of werewolves were also common in Eastern Europe. A good example is a Lithuanian tale recorded by Edmund Veckenstedt of a peasant who repeatedly went to the stable in the morning and found the mangled bodies of his horses with their necks bitten through. One night, he stayed up to watch the stable, and his neighbor approached carrying a bundle of sticks. The intruder threw the sticks to the ground and began to roll over them. After passing the first stick, his head became that of a wolf, and with every subsequent stick another part of his body changed, until finally he was entirely a wolf. In this form, the neighbor raided the stable, but, while he was gone, the peasant quickly took one stick away. After his predations were finished, the wolf rolled over the sticks once more, and changed, part by part, back into a man. But the transformation was not complete. The peasant accused his neighbor in town the next morning, and a physical examination showed that the culprit had the tail of a wolf.

The most famous literary product of this period is the story of "Little Red Riding Hood," who is lured from the path to her grandmother's house by a wolf. Early versions of the story, including the version in *Stories of Mother Goose* by Charles Perrault (1697), are all simple warning tales. Perrault ended his tale as the wolf, having already devoured the grandmother, ate up the little girl. He concluded with the moral:

> Wolves may lurk in every guise.
> Handsome they may be and kind,
> Gay and charming—never mind!
> Now, as then, tis' simple truth—
> Sweetest tongue has sharpest tooth!

Perrault clearly wanted to drive home the lesson as emphatically as possible, but many people could not bear the harshness of the ending. In the somewhat convoluted version from the collection of fairy tales published by the Grimm brothers, a woodsman rescued Little Red Riding Hood and her grandmother from the belly of the wolf.

In the heavily populated regions of Northwest Europe, wolves were less of a threat, and a more complex, if not always more favorable, view of them emerged. As the antique tradition of the beast fable was developed in Medieval Europe, it was increasingly adapted to social commentary and various sorts of animals were used to represent different classes. The wolf served as a satiric portrait of a monk. The relatively long hair covering the heads and bodies of wolves reminded storytellers of the robe and cowl of a monk. Even more significantly, monks and wolves shared a reputation for greed. Finally, in the middle of the twelfth century, an unknown Flemish author wrote a humorous epic about the wolf-monk entitled *Ysengrimus*. The monk, Ysengrim, ingeniously tried to reconcile his ravenous appetite for sheep with monastic law, and he claimed to practice a "religion of the stomach." He matched wits with the peasants, and with his companion the fox, until a clan of swine finally killed him.

Over the next few centuries, the wolf Ysengrim played a prominent role in the cycle of Renard the fox, who emerged as a cunning peasant. No longer a monk, Ysengrim had become a naïve noble who fell prey to the wiles of his small but unscrupulous adversary. He was constantly beaten, cuckolded, and maimed, and the stories about him left readers torn between frustration, laughter, and pity. He represented the aristocratic order that was slowly but inexorably starting to give way to the emerging bourgeoisie.

Aristocratic houses had often adopted the wolf as their heraldic symbol; as regimes became increasingly democratic, the wolf was blamed for the cruelty and intemperance of royalty and the nobility. This was particularly true in the United States, where the campaign to eliminate predators such as the wolf became a moral crusade. By the 1920s the grey wolf, which once had probably numbered in the millions, was almost completely extinct in the United States.

But the disappearance of the wolf created a wave of nostalgia. At the end of the nineteenth century Rudyard Kipling, known as "the poet of the British Empire," wrote *The Jungle Book*, a collection of animal stories centered mostly on an Indian boy named Mowgli who was raised by wolves. Like Britain, the wolves have both a king and a parliament, but the "law of the jungle" which they follow is sterner than that of a modern republic. The den of wolves is essentially a military school, which endeavored to make aristocratic young boys into "men."

As World War I began to approach, ever more people felt admiration for the martial qualities of the wolves. German authors such as Hermann Hesse and Hermann Löns wrote stories in which the wolf appeared as a representative of a natural order, heroically resisting the encroachments of civilization. In a story entitled "Lobo, the King of the Currumpaw," (1900) the Canadian-American author Ernest Seton-Thompson celebrated a wolf named Lobo as a romantic outlaw.

The regime in Nazi Germany, seeking to cultivate the fierceness and cruelty often associated with wolves, made the wolf into a sort of cult. Adolf Hitler himself adopted the nickname "little wolf," and he gave his various headquarters names like "Wolf's Lair," "Wolf's Gulch," and "Werewolf." He once explained that people cheered him rapturously because "a wolf is born." Near the end of World War II, a Nazi commando unit formed to operate behind enemy lines was called the "Werewolves." On a more practical level, the Nazis introduced the first legislation for the protection of wolves.

The United States and other countries, however, were not very far behind. Aldo Leopold in his essay "Thinking like a Mountain" (1948) tells how he, as a young man, had never passed up an opportunity to kill a wolf. One when day he shot an old wolf and her pups on a mountainside. This passage, which expresses regret for the deed, is perhaps the most famous in all of American nature writing:

> We reached the old wolf in time to watch a fierce green fire dying in her eyes. I realized then, and have known ever since, that there was something new to me in those eyes—something known only to her and to the mountain. I was young then, and full of trigger-itch; I thought that because fewer wolves meant more deer, that no wolves would mean a hunters' paradise. But after seeing the green fire die, I sensed that neither the wolf nor the mountain agreed with such a view.

At about that time, American conservationists under the leadership of Leopold began to defend wolves as an integral part of the environment.

With government protection, the population of wolves in America began to rebound in the latter half of the twentieth century. In 1963, naturalist Farley Mowat published a fictionalized account of his sojourn in the wilderness of Alaska entitled *Never Cry Wolf*, in which he wrote about a pair of alpha wolves named George and Angela. The book quickly became a best-seller and established a popular image of wolves as living an ideal of family life. Mowat's praise of the marital fidelity of wolves resonated in a society where people were increasingly concerned about divorce. In the 1990's,

Clarissa Pinkola Estes published her bestselling *Women Who Run with the Wolves*, in which she used the wolf to symbolize the "wild woman archetype." That is the female who, though nurturing to husband and children, retains a primal connection with the natural world.

Today, in America and much of the world, a craze for wolves continues, and they are often seen on posters and in jewelry. Nevertheless, the old ambivalence towards wolves remains, and the old hatred of them can erupt unexpectedly. Plans to reintroduce wolves in Yellowstone National Park were implemented in the 1990s over vehement opposition from ranchers.

7
MERMAID'S COMPANIONS

THOUGH THE SURFACE OF THE EARTH HAS LONG BEEN THOROUGHLY explored, the floor of the ocean retains, at least for now, most of its ancient mystery. Mythologies anticipated modern science by having life emerge from deep waters. Tradition makes the ocean a remnant of the primeval chaos, surrounding the land and its kingdoms. It impressed early people as an endlessly fertile womb, from which new forms of life constantly emerged. The variety of life within the ocean is far greater than that on land. The structural similarity among most vertebrates, with their legs and arms, is easy to observe. Many invertebrates from the sea, from octopuses to clams to sea anemones, seem entirely unlike anything else in the entire world. Within the ocean, it is far harder to tell animals from plants, or life from death. Where the water is so deep that no sun can enter, most creatures glow from within. This phenomenon, known as "bioluminescence" is confined to very few forms of life on land, such as mushrooms and fireflies, but in the ocean it is shared by jellyfish, octopuses, squids, sea stars, clams, anemones, fish and many other animals. They carry their own light, a bit like heavenly bodies and saints.

CLAM, OCTOPUS, SEA STAR, SQUID, AND CRAB

The Kraken sleepeth: faintest sunlights flee
About his shadowy sides: above him swell
Huge sponges of millennial growth and height;
And far away into the sickly light,
From many a wondrous grot and secret cell
Unnumber'd and enormous polypi
Winnow with giant fins the slumbering green.

—ALFRED LORD TENNYSON, "The Kraken"

According to Hesiod's *Theogony*, Aphrodite, the Greek goddess of love, first emerged from the ocean when it was fertilized by sperm of the castrated god of the sky, Uranus. Aphrodite was sometimes portrayed as floating towards the shore on a scallop shell, after being created from the foaming sea. Shells, and vessels modeled after them, were later used as baptismal fonts in Christianity, and they came to be associated with the Virgin Mary, who took over many attributes of the pagan goddesses that preceded her. The starfish is known in Latin as the *Stella Maris* or "star of the sea," taking its name from the Virgin Mary in her capacity as a guardian of mariners. In secular culture, it is a positive symbol of the richness and bounty of the sea. Just as mariners navigated by the stars, starfish are often shown pointing directions with their arms.

In many cultures, shells have been used as a medium of exchange. Wampum, strings of clamshells fashioned into beads, was an object of ceremonial exchanges originally used by American Indians of the eastern woodlands. Since the Bank of England did not allow the American Colonies to coin their own money and European currency was in short supply, the Colonists often adopted wampum for their own transactions, and in 1761 they even set up a factory in New Jersey to manufacture wampum. The clam provides a more vivid symbol in death than in life, for living clams are often thought of as slimy. On the other hand, the phrase "happy as a clam" designates a person who leads a thoughtless, carefree existence.

The octopus, according to recent research, is exceptionally intelligent, and it also shows an emotional expressiveness that is perhaps (at least from

Illustration from mid-century America showing a man being assaulted by an octopus.

a human point of view) unique among invertebrates. Not only does it seem to express moods through changes in color, but its gestures also appear remarkably articulate. As might be expected, people have been disconcerted as well as charmed by these signs of intellect and emotion. The octopus is a common motif on jars and artifacts from Crete and other early Aegean civilizations, where its limbs are often stylized in curvilinear patterns. In Christian culture, its squirting of black ink to blind other creatures has made it a symbol of the Devil, while its many arms have sometimes made it stand for lechery. It is often depicted on chests of a sunken treasure, as a guardian of the watery depths.

In 2010, Paul the Octopus, in The Sea Life Center in Oberhausen, Germany, became a huge international celebrity, by successfully predicting the outcomes of all of the German national football team's seven matches, plus the World Cup final football match between the Netherlands and Spain in 2010. The predictions were arranged by placing Paul between two bowls of mussels representing each of the opposing teams, and the bowl he chose was taken to represent his selection of the winner. Statisticians calculated that the chances of his success being due to chance were miniscule. Was Paul really psychic? If not, the celebrity he achieved becomes an even bigger testimony to his charisma. A vast number of animals are now being employed to predict sports events (pythons, cows, otters, elephants and others), but none has quite managed to equal Paul's record or his renown.

Closely related to, and often confused with, the octopus is the squid, though it does not have the same simplicity of form. The squid tends to be more aggressive and, in addition to the eight arms of the octopus, it has two tentacles that are used for seizing prey. A giant squid, so large it could pull down boats, was reported in the sixteenth century by the Swedish naturalist Olaus Magnus and by many explorers over the next several centuries. Superstitious mariners scored a remarkable triumph over skeptical scientists when several enormous squids, one about 55 feet long, were washed ashore on the coast of Norway in the 1870s. This creature may have been the prototype of many ancient monsters of legend such as the Greek Scylla who seized men from the ship of Odysseus.

The ocean mirrors the sky, and many creatures of the watery depths, such of the octopus and lobster, have sometimes been observed in constellations. The crab, "Cancer," was sent by Hera to harass Hercules as he battled the seven-headed hydra. The hero, brandishing a torch in one hand and a sword in the other, still managed to crush the crab under his feet. In Japan, members of a species known as "samurai crabs" (*Heikeopsis japonica*) are said to be the spirits of the Heike warriors who committed suicide by throwing themselves into the sea after losing the naval battle of Dan no Una to the Genji clan. Each crab, according to legend, bears the face of a drowned warrior on its back.

All of these sea creatures are more-or-less anomalies, unlike anything else in the natural world. Perhaps human beings, feel fascinated by them because we ourselves are an anomaly, constantly troubled by our isolation from the rest of nature. Many of these creatures seem to show a glimmer of the human spirit in a paradoxical way. The crab, with its zigzag walk, seems to reflect the human propensity towards hesitation. The remains of these creatures are a tangible link with a mysterious kingdom, and so, in the nineteenth century, relics such as shells became favorite souvenirs for tourists visiting coastal towns.

DOLPHIN AND SEAL

That dolphin-torn, that gong-tormented sea.

—W. B. YEATS, "Byzantium"

Dolphins and seals are aquatic mammals, and both seem to have a special affinity for human beings. Seals spend a good deal of time sunning themselves on coastal rocks, staring at people in the distance. They will usually swim away if somebody approaches, though in heavily populated coasts they sometimes lose their fear of human beings.

Seals, thoroughly comfortable in water, are endearingly awkward on land, but folklorist David Thompson writes that no other animal, "not even the hare, has such a dream-like effect on the human mind . . . " The American poet Hart Crane refers, at the end of his poem "Voyages," to "the seal's wide spindrift gaze toward paradise." Seals observe human dwellings with apparent wonder, similar to what people living along coasts often feel for the sea, and some legends maintain that seals are the spirits of drowned men and women. According to Celtic lore, the one-eyed giant Balor, king

The two seals have very human expressions. The male gazes at a ship, suggesting a kinship with human beings.

(From a British book on natural history, c. 1850)

This mermaid and merman are depicted with a lot of realistic detail, but they are composites of many creatures on both land and sea.

(From Historia Monstrorum *by Ulisse Aldrovandi, 1642.)*

of the Fomorians, once heard a prophecy that he would die at the hands of his grandson, and so, to avoid that fate, he ordered that the children of his daughter Ethlinn be thrown into the ocean. One of them survived, returned to vanquish his grandfather, and became the god Lugh, while the others turned into seals.

Mermaids and mermen have been part of myth and legend since at least ancient Babylon, where they were depicted on walls. Many folklorists believe that the legends of the mermaid first originated from observation of seals. Others, however, believe it originated from manatees, relatives of seals, that appear far less human in form but more so in their locomotion. At any rate, men have often interpreted the gaze of female seals as one of amorous longing. The selkies are figures of British and Irish folklore — seals that slough off their coats to become human and dance together for an evening on the shore. Sometimes a man will manage to steal the skin of a selkie maiden and win her as a bride, though she will generally find the skin, resume the form of a seal, and rejoin her herd. Many families along the northern and western coasts of Britain and Ireland trace their ancestry to seal people, and people in some families can even show webbed hands to prove their origin. According to many traditions, land animals such as cows and sheep all have an equivalent in the water, and, along the coasts of Scotland, seals are considered the "people of the sea."

In the area around the Mediterranean, that status goes to dolphins. Dolphins are confined to water, but they follow ships at sea, often leaping into the air. There are many stories, some probably true, of dolphins rescuing people from drowning. People have long understood the upturned mouth of the dolphin as fixed in a perpetual smile. They have been particularly beloved in Greece, where they were sacred to Poseidon, god of the sea. They often drew his chariot, and accompanied the nereids and tritons in his entourage. Pliny the Elder wrote that, "Dolphins are not afraid of humans as something alien but come to meet vessels at sea and play and leap around them; they try to race ships and overtake them even when they are in full sail."

Apollo was known as "Delphinius" or "Lord of the Dolphin." In the Homeric Hymn "To Pythian Apollo," the god of the sun had been looking for priests for his temple, when he caught sight of a boat containing pirates from Crete. Apollo changed himself into the form of a giant dolphin and leapt from the sea into the ship. The sailors were too frightened to lower their sail, and a wind drove their ship until it reached the shine at Delphi. Then Apollo took the form of a young man and told the pirates that he had brought them there to be keepers of his sanctuary.

The dolphin was also sacred to Dionysus, a sort of shadow-image of Apollo, who shared the shrine at Delphi with him. Apollodorus, Ovid, and others tell how the god once chartered a pirate ship to the island of Naxos, not far from Crete, which was to be a center of his worship. The pirates, not realizing Dionysus was a god, sailed past his destination, thinking to sell him as a slave. The mast and the oars suddenly turned into snakes, the craft filled with ivy, panthers appeared, and the wild music of flutes drove the pirates mad. The men leapt from the boat in terror and became dolphins.

In part because their leaps are so rhythmic, dolphins are reputed to be fond of music. In a popular story by Herodotus, Arion, a peerless harpist, once hired a boat to take him to Corinth. Once they were on the open sea, the sailors decided to kill Arion and steal his wealth. Arion entreated his captors to allow him to sing for one last time on the quarterdeck, after which he promised to end his life. The sailors, delighted at the opportunity to hear

A monkey, having heard the story of Arion, impersonates a human being and rides on the back of a dolphin in this Aesopian fable.

(Illustrated by J. J. Grandville from Fables de la Fontaine, *1839)*

his song, quickly agreed. When he had finished his song, Arion jumped from the boat and landed upon the back of a dolphin, which then took him to his destination. Christians would later think of Arion as a martyr, while the dolphin that bore him away became a symbol of Christ the savior.

The salvation of Arion was commemorated at Corinth with a bronze statue of a man riding upon a dolphin, and this became a popular motif in Greco-Roman art. Since very early times, the dolphin has been a favorite symbol of port cities in heraldry and on coins. Roman coins since the time of the Emperor Titus showed a dolphin, considered the fastest of animals, entwined around an anchor with the motto, *Festina lente* or "Make haste slowly."

The lore of the dolphin belongs mostly to maritime culture, in which the ways of many nations have always blended. Because they seem to have such an affinity for human beings, sailors have often watched the motions of dolphins in order to forecast weather and, occasionally, fortunes in war. In *The Tale of the Heike*, a thirteen century CE Japanese epic of the war between the mighty Heike and Genji clans, a school of about two thousand dolphins suddenly appeared before the decisive sea battle of Dan-no-ura. "The Genji will be destroyed if the dolphins stay on surface and then turn back; we will be endangered if they dive and pass us," a Heike oracle predicted. As soon as these words were out, the dolphins immediately passed under the Heike boats, and the commanders realized that they were doomed.

Occasionally, however, people have interpreted the attraction dolphins appear to feel for human beings as a longing for a tragically departed love. The sounds of the dolphins as they play become a bittersweet lament. In Indonesians folklore, a man once beat his wife for giving one of his fish to her son. She went down to the sea to wash off the blood and found herself changed from the waist down into a dolphin. Her husband later changed his ways, set out in search of his wife, and finally was transformed into a porpoise, but the two were never reunited.

With the use of larger, more mechanized ships, traditional maritime culture has changed, and even sailors probably no longer feel as intimate with the sea as they did before the Industrial Revolution. As people have increasingly grown to value mental over physical abilities, however, there has been a revival of interest in dolphins. They have long held a reputation as being among the smartest of animals. Popular writers and some scientists of the twentieth century CE have speculated that they might have a sophisticated oral culture to rival that of humankind. A few science fiction writers have even wondered if they might eventually challenge human supremacy.

But loneliness, far more than intellect, draws us to seals or dolphins and, perhaps, interests them in us. For all their well-deserved reputation for drinking, brawling, and whoring, mariners traditionally led a very austere life most of the time. Almost alone on the wide sea, any miracle of love could seem possible.

8

THE BARNYARD

TRADITIONS RUNNING FROM THE LEGENDARY AESOP TO CHAUCER AND beyond make the barnyard a place of merriment and adventure. At least by comparison with their twenty-first century counterparts, the animals in those barnyards seem a lot more "wild" than "domestic." Death by human hands remains largely impersonal, since, for the animals, human beings are largely avatars of fate. Until the time for that has come, the denizens of the barnyard must constantly match their wits against foxes and other predators. The fence separating the barnyard from adjacent forests and fields is the final boundary between civilization and the wild, and so farm animals are, in many ways, the first line of defense. Though they will kill and eat the animals in time, farmers are very protective of them; the predations of carnivores can arouse a truly furious human response. Animals of the barnyard are especially stylized to reflect mores of human society. The rooster is a protective head of the household, while the hens are very maternal. As we approach contemporary times, however, the freedom of barnyard animals gradually diminishes. In E. B. White's novella *Charlotte's Web* (1952), they continue to have lively adventures, though these are shadowed by a constant awareness of mortality. In George Orwell's *Animal Farm* (1945), from the same era, however, the barnyard becomes a metaphor for the totalitarian state. In the twenty-first century, people are reviving the practice of raising animals, especially chickens, in suburban and yards and even on urban rooftops.

BULL AND COW

Man in his prosperity forfeits intelligence.
He is the one with the cattle doomed to slaughter.

—PSALMS 49:20

Bulls and cows already are prominent in the Paleolithic paintings on the walls of caves in France, Spain, and other parts of Europe. In the main chamber of the cave at Lascaux, five enormous bulls decorate the ceiling. In the homes of Çatal Huyuk (near Jericho, in Turkey), are many shrines to bulls from the middle of the ninth millennium BCE. Large bulls' heads of modeled in clay extend from the walls. Similar shrines to bulls have been found in much of the Mediterranean area. Only very slowly did people lose their fear of these giants. Cattle were not domesticated until around 3,000 BCE in Europe, long after other animals such as the dog, sheep, and goat.

There is a very intimate association of man and bull in the religion of Zoroaster, where sacred texts tell that Ohrmazd made a lone white bull, "shining like the moon," as the fifth act of creation. He then made the first man Gayomart, as the sixth. The seed of man and bull were then created from "light and the freshness of sky," so that both would have abundant progeny. As the world draws to an end, Soshyant, a descendant of Zoroaster, is to sacrifice a great bull named Hadhayans, and its fat will be used to make the elixir of eternal life.

As the largest of domestic animals, the bull was the supreme sacrificial offering throughout almost all of the ancient Mediterranean. The skin, bones, gristle, and a small bit of meat were left on the altar for a god, while people feasted on the rest. Some, however, thought it impious to give the gods such a tiny share. On important occasions, the Hebrews would perform a "holocaust," a sacrifice in which the entire animal was offered up to God. The Bible gives a very detailed description of the bull sacrifice that accompanied the investiture of priests. Some blood was placed around the bull's horns by the altar to purify it, and the rest was poured out onto the ground. Every part is of the bull was disposed of according to a precise ritual (Leviticus 8:14-17).

Greek painting of a bull being led to the altar for slaughter.

In Greece the sacrifice of a bull was generally reserved for Zeus, in Rome for his counterpart, Jupiter. The slaying of the bull became the central rite in the religion of Mithras, which rivaled Christianity in popularity during the latter part of the Roman Empire. Mithras, accompanied by a dog and other animals, would plunge his sword into a great bull at the end of the world, so that all things might live again. Artists of the Roman Empire would show grain sprouting from the wounds of the bull as it was slain by Mithras.

In one myth of the Greeks, Poseidon, god of the sea, had given King Minos of Crete an enormous bull, intending that Minos should offer it back as a sacrifice, but Minos kept the bull for himself instead. This act, a possible allusion to the first domestication of animals, began a sequence of events in which great buildings were erected, unnatural acts performed, and people sacrificed. The angry god caused the king's wife, Pasiphaë, to fall in love with the bull. She ordered Daedalus, the great inventor, to construct a hollow cow of wood, and she crept inside. From there, she made love to the bull and conceived the minotaur, a monster with the head of a bull and the body of a man. Deeply ashamed, Minos ordered Daedalus to construct a labyrinth, an underground series of passageways, to house the minotaur. When his son Androgeos was killed by Athenians, Minos demanded that the Athenians send a tribute of seven youths and seven maidens every year as penance. The young people were placed in the labyrinth, where they wandered until they were either eaten by the monster or else died of hunger themselves. Theseus, a prince of Athens, volunteered to go as one of the youths. Ariadne, the daughter of Minos, fell in love with him. She gave him a ball of yarn to unroll as he wandered through the ~~minotaur,~~ *labyrinth*, as well as a sword to do battle with the creature. Theseus killed the minotaur, returned to the entrance following the trail of yarn, and escaped Crete with Ariadne.

The tale mocks the rulers and the religion, especially fertility rites, of the Cretans, often adversaries of the Greeks. Minos, a fool and tyrant in Greek mythology, appears to have been an actual ruler in Crete. He claimed the bull as an ancestor, and he did indeed have an elaborate palace with underground chambers at Knossos. The minotaur was a Cretan god, whom the Greeks considered a grotesque, unnatural creature. He anticipated the Christian devil with his horns and his dwelling beneath the ground.

Ancient Cretan wall paintings showed acrobats turning somersaults on the back of a bull. In the Mesopotamian tale of Gilgamesh, our earliest heroic epic, the bull of heaven represents a terrible drought, which came early in the second millennium and may have destroyed the mighty Akkadian Empire. The bull was sent down to earth as a punishment because Gilgamesh and his companion Enkidu had cut down the cedar forests of Lebanon and killed their guardian, Humbaba. The bull of heaven immediately killed hundreds of people. Enkidu grabbed the horns of the bull and leapt aside, and then Gilgamesh killed the bull with his sword. Despite this tale, the worship of the bull persisted in Mesopotamia, where people identified it with Anu, the god of the sky, and Adad, the god of storms. There are a number of depictions of bull-men in Mesopotamian art, and scholars speculate that some may represent Enkidu.

In Egypt, the Apis bull was considered an incarnation of the creator god Ptah, conceived when fire came down from heaven and impregnated a cow. Priests identified this bull by searching among calves for one with very specific markings. He would be black, but with an inverted white triangle on his brow. A mark shaped like the silhouette of a vulture stretched across

Assyrian winged bull from the palace of Sargon

his shoulders, and there would be a crescent moon on his sides. He would also possess a sign like a beetle on his tongue. Once the Apis bull was found, there would be great rejoicing. He would march in a great procession; women would pray to him for children, and priests would perform sacred rites. Finally, he would be brought to the temple of Ptah, where he would live. When he finally died, the bull would be consecrated to Osiris, god of the dead, and buried in great splendor. Then priests would search for his successor.

The Greeks and Romans, who had anthropomorphic deities, sometimes thought the animal-gods of Egypt were strange or primitive. Nevertheless, they generally respected the Apis bull. Herodotus told how Cambyses, King of the Persians, committed a sacrilege against Apis. After conquering Egypt, Cambyses arrogantly threw his dagger at the sacred bull, striking the animal in the thigh. Then Cambyses ordered the priests of Apis to be beaten and forbade them to celebrate any festivals under penalty of death. The Apis bull died unattended in the temple and was secretly buried. Shortly afterwards, the gods struck Cambyses with madness. He killed his

Illustration by Albrecht Dürer to Sebastian Brandt's Ship of Fools *(1494), showing worship of the golden calf.*

brother, sister, and many trusted servants in fits of temper. Finally, after he had driven his subjects to revolt, Cambyses accidentally wounded himself in the thigh with his own sword. When he realized that the place of the wound corresponded exactly to that where he had struck the Apis bull, the horrified Cambyses knew that he was doomed. The wounded limb became infected with gangrene and he died shortly afterwards.

The golden calf, which the Israelites worshipped on the flight from Egypt, is a form of the Apis bull. The Hebrews were constantly struggling with the old animal cults, and Moses put this one down with great ruthlessness, killing about 2,000 people. Today, bullfights reenact the struggle against the ancient cult of the bull. It is remarkable that, with all the power of our technology, we still need such ritual affirmations of human dominance.

The many deities of the ancient world who were depicted in the form of a bull also included the Greek Dionysus, Phoenician Moloch, and the Syrian Attis. Siva, the Hindu god of creativity, rides on a white bull, and so the cow and bull are sacred in much of India. The cow and bull were also important in the culture of the Far East, but there they have been viewed with a bit less awe and more intimacy. The ox is a sign in the Chinese zodiac. The sage Lao Tzu has often been depicted riding on a water buffalo, and so are small boys playing the flute. Cattle in Asia are not only admired but often loved as well.

With many animals such as the cat and dog, worship eventually led to domestication. In a similar way, religious awe can evolve into economic power. The value of coins, initially, was measured according to the animals they might buy, and the earliest coins are stamped with pictures of cattle. In Homeric times, a coin known as "a talent of gold" was the equivalent of an ox. Our word "pecuniary" comes from the Latin "pecus," meaning "domestic herd animal." Expressions such as the "growth" of investments also hark back to a time when herds and flocks were the major measure of wealth.

It is strange that English has no common word that can stand for either "bull" or "cow" in the singular. The closest, perhaps, is "bovine" which sounds a bit pedantic. When we speak of a "rabbit" or a "spider," the subject may be either female or male. For the cow and bull, however, sexuality is such an intimate part of their identity that, apparently, even the word cannot dispense with gender. The same is true in other languages such as Latin. Both bull and cow are extremely important in the religious history of humanity, but their symbolism is so different that they sometimes hardly seem to be of the same species. As we have seen, the bull is generally associated with generative power and energy. The cow is, by contrast, maternal.

The cow has often been worshipped as a provider of milk, and, unlike the bull, she was seldom sacrificed. In Norse mythology, the creation of the world begins when the cow Audhumbla was created from melting frost. She nourished the giant Ymir with rivers of milk. For nourishment, Audhumbla licked salt on ice of the frozen waste, gradually uncovering a human figure—Buri, ancestor of the gods.

The Egyptian goddess Hathor, mistress of the underworld, was often depicted in the form of a cow. As goddess of love, music, and fertility, she was among the most beloved of deities, though she could become vengeful and extremely dangerous. In Greek mythology, Hera, the wife of Zeus, is referred to as "cow-eyed" by Homer, and may have been a bovine deity in very remote times. Her husband Zeus constantly had affairs with mortals. In one tale, Hera surprised Zeus when he was making love to the maiden Io. Hoping to cover up his transgression, Zeus turned Io into a heifer. Hera saw through the trick, and sent a fly to torment the transformed maiden, driving the poor victim all over the world.

Hollow images of cows inside of which people might be buried were made in Egypt and elsewhere in the ancient world. Herodotus told how the daughter of the pharaoh Mycerinus, in her last moments, asked her father if she might still see the sun once a year. Devastated by her death, Mycerinus ordered a large cow to be made of wood, hollowed out, and covered up with gold. A golden orb representing the sun was then placed between the horns, and the young woman was buried in the cow. Lamps were kept always burning beside the cow, and once a year the cow was raised and exposed to the light of day.

What is greatly valued easily becomes an object of contention. Cattle raids were frequent and might even lead to wars. In Greek mythology, the infant Hermes, who later became messenger of the gods, stole the cattle of the sun god Apollo. He killed two and locked the rest in a cave. After eating them, he strung the guts of the cattle across a tortoise shell to make the first stringed instrument, a lyre. Apollo, who could tell the future, easily found the thief. The mother of Hermes protested that her son was but an infant and certainly not capable of theft, but Apollo demanded that his property be returned. Finally, Apollo exchanged the cattle for the lyre and thus became the patron of music. At the time, wealth was measured far more often in the form of cattle than of money. For the very rich, from ancient Greece to the present, wealth has often been measured by possession of art, and tycoons such as Mellon, Carnegie, Morgan, Rockefeller, and Getty have often been avid collectors. The story of the trade between Hermes and Apollo contrasts

Some of the forms taken by the Egyptian cow-goddess Hathor.

these two forms of wealth, the more tangible form of livestock and the more spiritual one of art.

Perhaps the most important epic of the Celtic people is the *Tain Bo Cuailnge* or "Cattle Raid of Cooley." As it begins, Queen Maeve and her husband Ailill were arguing about which of them brought greater wealth to their marriage. They compared their goods, which included clothes, jewels, and sheep. Maeve was able to match every possession of Ailill except for a white horned bull named Fennbhennach. There was only one other bull so fine in all of Ireland, a bull named Donn Cuailnge in the province of Ulster. Queen Maeve resolved to have that bull, and, when it was refused to her, she sent an army to steal it. The hero Cu Cuchulainn valiantly defended Ulster, but after many battles, Maeve finally carried off her prize. The story ended with a fight between the two bulls, and Donn Cuailinge was victorious. It wandered about with the entrails of Fennbhennach on his horns, and then finally died of its wounds. The story is not altogether unlike many American westerns, once as popular as cheap paperbacks and early television shows, in which "cattle rustlin'" often ignites a "range war."

Even after millennia of domestication, cattle have not entirely lost their numinous quality. Those who are associated with cattle seem to gain some of their power and virility, at least in popular imagination. This is so for the heroes of ancient epics as well as the gauchos of Argentina and the cowboys of the American West. Eating beef is still associated with strength and virility, so muscular men are referred to as "beefcake."

Ancestral patterns can be very persistent, even when the culture no longer appears to sanction them. Despite the Christian ban on animal sacrifice, Medieval bestiaries often saw a symbol of Christ in the ox or bull that is slain for food. In 1522, desperate to stop the black plague, Pope Leo X allowed bulls to be sacrificed in the Old Roman Coliseum, though to no avail. On farms in England, sacrifices of bulls have been practiced from time to time, sometimes even in the twentieth century CE, for such purposes as stopping disease or witchcraft. The bulls have sometimes been killed in very brutal ways, such as being buried upside down or burned alive, when people thought that was what the magic required.

When Holy Roman Emperor Charles V had a son, later to become Philip II of Spain, he celebrated by publicly killing a bull in the marketplace. The most notable survival of animal sacrifice may be in the bullfights of the Iberian Peninsula and Latin America. The popularity of bullfighting began at start of the modern era, around the sixteenth century CE. The elaborate ceremony and pageantry now associated with bullfighting date only from the late 1800s. The bull is systematically enraged by being kept in darkness, and then abruptly exposed to the bright lights of the arena. Lancers on horseback systematically goad the bull, then, when the animal has been worn out, the matador delivers the fatal thrust with his sword. The triumph of the matador's finesse and skill over brute power symbolizes the victory of "civilization" over "barbarism." Despite the obvious cruelty, matadors insist that they respect and even love the bulls. On a barely conscious level, the appeal of the bullfight may, like archaic sacrifices, be based on the expectation that death can release cosmic energy, which then nourishes all of life.

COCK AND HEN

It faded on the crowing of the cock.
Some say that ever, 'gainst that season comes
Wherein our Savior's birth is celebrated,
This bird of dawning singeth all night long;
And then, they say, no spirit dare stir abroad,
The nights are wholesome, then no planets strike,
No fairy takes, nor witch hath power to charm,
So hallow'd and so gracious is the time.

— WILLIAM SHAKESPEARE, *Hamlet act 1, scene 1*

Aelian wrote, in the second century CE, of two Greek temples separated by a river, one consecrated to Hercules and the other to his wife, Hebe. Cocks were kept in the temple of the god and hens in that of the goddess. The roosters would cross the waters once a year to mate, returning with any male offspring and leaving the females for the hens to raise. The arrangement is not very plausible, among other reasons because cocks generally cannot stay together without fighting, and so barnyards have only a single rooster per flock. Nevertheless, the account shows how cock and hen, even more than other animals, seem to be defined by their gender, to a

Two cocks fighting over a hen.

(Illustration by J. J. Grandville—from Fables de la Fontaine, *1839)*

155

point where they hardly appear to belong to a single species. Both cock and hen were indeed kept for sacrifice in temples throughout the ancient world from Egypt to Greece. On the altars, their entrails were used to predict the future.

Until recent times, even urban dwellers would generally be woken by the call of a rooster at dawn. From ancient times through much of the Middle Ages, the crowing of the cock at certain times was so predictable that it was used to signal the changing of the guard. It had a triumphant ring, and was said to frighten away the spirits of darkness. The crowing of a cock served as the voice of conscience in the Bible after Peter had denied knowing Jesus, since the sound moved him to tears of regret (Matthew 26:75). The red comb of a cock heightened its association with the sun. Roosters themselves have always been celebrated for their fierceness, as they seemed to lord over the barnyard.

The cock is also a solar animal in East Asia, where it is the tenth sign of the Chinese zodiac. According to one Japanese tale, the sun-goddess Amaterasu, angry at the violence of the storm-god, moodily withdrew into a cave, leaving the world in darkness. When a cock crowed, she wondered if the dawn had come without her and went to the entrance of her cavern, to see that indeed it was bright day.

Hens, by contrast, are symbols of domesticity and maternal care. Especially when brooding on their eggs, they seem unconcerned about all else, even the cock. In the Bible, Jesus says, "Jerusalem, Jerusalem . . . How often have I longed to gather your children, as a hen gathers her chicks under her wings, and you refused!" (Matthew 23:37).

Long before Christ, the cock symbolized resurrection. The cock was associated with Asclepius, the Greek god of healing, who as a mortal physician once raised a man from the dead. The last words of Socrates, as recorded in Plato's dialogue "Phaedo," are, "Crito, we ought to offer a cock to Asclepius. See to it and don't forget." Perhaps Socrates wished to thank the god for spiritual healing as he moved on to the next world.

Because cockfighting is extremely ancient, widespread, and practiced virtually wherever roosters are found, many anthropologists believe that the birds were initially domesticated for the sport rather than for meat. Their willingness to fight one another to the death has made cocks a symbol of the warrior spirit. Before such battles as Marathon and Salamis, Greek commanders would rouse their men to battle by showing them fighting cocks. The general Themistocles ordered an annual cockfight in Athens to commemorate the victory of the Greek over the Persians. Fighting cocks might also be used to predict the outcome of a battle. In the Medieval Japanese *Tale of the Heike*, a local warlord uses a cockfight to decide which side to take in the war between the Heike and Genji clans. He matched seven cocks that were white, the color of the Genji, against seven that were red, the color of the Heike. When all of the white cocks had proved victorious, he knew that the Genji would also win.

At the start of the seventeenth century, the naturalist Ulisse Aldrovandi of Bologna wrote of two friends that sat down to eat a roasted rooster. When one casually remarked that even Christ could not raise the bird from the dead, the cock immediately leapt up, splashing the men with sauce, and then turned them into lepers. In a version of the story from Ireland, a group of unbelievers was sitting around a fire over which a cock had been boiled. "We have buried Christ now," said one "and he has no more power to rise from the dead than the cock in this pot." Immediately, the cock rose and crowed three times, saying, "The Virgin's son is saved." Versions of the story are set to music in "King Pharim" and other Christmas carols.

The cock also experiences a sort of resurrection in a famous story from the work of Alcuin, a learned monk at the court of Charlemagne. A rooster, boasting of his powers, forgot to remain watchful and suddenly found the jaws of a wolf had closed about his neck. The cock begged to hear the wolf sing just once, so he would not have to die without hearing the wonderful harmonies of a lupine voice. The wolf opened his mouth to grant the request, at which the cock immediately flew up to a tree and admonished the wolf, saying. ". . . Whoever is taken in by false pride will go without food . . . " For some Medieval readers, the jaws of the wolf represented the grave or, perhaps, the gate of Hell, and the bird was saved not only by his cleverness but also by grace. In later versions, the adver-

The cock outsmarts the fox in this medieval fable, retold by La Fontaine.

(Illustration by J. J. Grandville from Fables de la Fontaine, *1839)*

sary of the cock was usually the fox, and the story—essentially a variant of Aesop's fable of "The Fox and the Crow"—has been retold by Marie de France, and countless other fabulists from the Middle Ages till the present.

The gently ironic version of the story by Geoffrey Chaucer in his *Canterbury Tales*, written in the late fourteenth century, elaborates the simple fable by adding heraldic symbolism and, with it, all of the pageantry of late Medieval courtly life. The cock, Chauntecleer, becomes a noble knight with seven hens as mistresses, one of which, Pertelote, has the special status of a full wife. They all speak in relatively formal verse, quoting not only the Bible but also classical authors such as Cicero and Cato, but, in the end, Chauntecleer is saved not by learning but by his quick wit. As Chaucer recognized, the moral that one should not trust flattery applies especially to powerful lords and ladies.

The use of the rooster as a symbol of France dates back at least to Medieval times, but it only became popular around 1789 during the French Revolution. This symbolism ultimately derives from a play on the Latin word "gallus," which can mean both rooster and Gaul. The Revolutionaries wished to break with their Royalist and Christian traditions, and the new French Republic was intended to be a "resurrection" of the nation's "gallic" past.

Since the cock and hen are so quintessentially male and female, people have often viewed any violation of their sexual roles as foolishness or worse. Using reasoning not entirely unlike that of modern sociobiologists, people in earlier eras often sought, in observation of animals, a sanction for customs such as female subordination. In the eighteenth-century Chinese novel *The Story of the Stone*, the aristocratic Lady Wang gently advises younger women to resigned acceptance of fate, quoting an "old rhyme":

> When rooster crows at break of day,
> All his hen-folk must obey.

According to a traditional belief that has been found from Germany to Persia, a hen that crows like a cock augurs terrible fortune and has to be killed immediately. A number of cocks were judicially condemned to death in the Middle Ages for laying eggs. Writing around the end of the twelfth century, Alexander of Neckam stated that an egg laid by an old cock and incubated by a toad could produce a cockatrice, a serpent able to kill with a glance.

Today, the proud society of the barnyard has largely disappeared, and most people rarely see fowl before it reaches the supermarket or the dinner plate, though heraldic roosters still decorate packages of cereal and many other products. Cock fighting is now illegal in the United States and most of Europe, but people, particularly from Latin America or the Caribbean, still practice it, believing they are preserving the values of a more heroic age.

GOAT AND SHEEP

> *I am not certain that anyone who has not spent time with shepherds can appreciate the intense involvement that exists between the shepherd and his flocks. The wellbeing of the flock is all, and everything else falls by the wayside. All that is done the whole year long is attuned to the single overriding consideration of the flock. Man's fierceness in defending his flocks and his lack of tolerance for anything he even imagines impinging on them are remarkable. To people who keep sheep, it almost seems, every other animal on earth could perish and it would be of no account.*
>
> —ROGER CARAS, *A Perfect Harmony*

Sheep and goats were, together with the dog, the first animals to be domesticated by human beings, around the end of the last ice age. Over the millennia, the symbolism and patterns of behavior these animals inspired have been especially intimately integrated into human culture. Sheep and goats are perhaps the only animals that have created not only an industry but also an entire way of life. Pastoral peoples must traditionally center almost every activity around their flocks, staying in one place for a time and then migrating when the edible vegetation is exhausted. Flocks inspire intense protectiveness, and they compel herders to constantly view predators such as wolves and even neighboring people as possible threats. By moving in unison and following a leader, sheep especially provide a model for understanding human society.

Sheep also provide wool, for clothing, and food. They will generally eat little besides grass, but they may be kept in rough, mountainous areas that are unsuitable for farming. Goats provide less meat, but they give copious quantities of milk. What is more, they can eat almost anything that grows, and they are even able to climb trees in order to get at their leaves. Tending flocks offered the most peaceful means to financial advancement in the relatively static societies of the ancient world, so shepherds were perhaps the first middle class.

The Biblical story of Cain and Abel records an early conflict between nomadic herders and settled agriculturists. Abel the shepherd offered the first born of his flock to God, while Cain the farmer offered his produce. God looked with favor only on the sacrifice of sheep. Consequently, Cain

God looking with favor on the sacrifice of Abel while Cain looks on in anger.

(Illustration from the mid nineteenth century by Julius Schnorr Carolsfeld)

killed his brother (Genesis 4:1-8), perhaps as a human sacrifice. Since herding requires a larger area of land than farming, growing population density gradually forced more people to turn to agriculture. Nevertheless, the vocation of shepherd remained an honored one in the ancient world, especially among urban dwellers who felt nostalgic for a simpler past. It offered many opportunities for solitude and contemplation. The Greek mythologist Hesiod tells in *Theogony* how he first became a poet when the Muses appeared to him as he tended sheep on the slopes of Mount Helicon.

The Egyptian god Amun was often portrayed with the head of a ram. The Greeks identified Amun with Zeus, who, they believed, had taken the form of a ram when the gods temporarily fled to Egypt in their war with the Titans. To commemorate his escape, Zeus later placed the ram in the zodiac, where it became the constellation Aries. Zeus was also identified with a domestic herd animal in his own right. When the goddess Rhea, his mother, hid the infant Zeus on the island of Crete to escape the wrath of his father Cronos, the fairy goat Amalthea suckled him. Later one horn of Amalthea broke off, and Zeus turned it into the cornucopia or "horn of plenty."

Perhaps because they seem to integrate themselves so well into otherwise forbidding landscapes, the Greeks constantly associated sheep and goats with flight and hiding. Odysseus, according to Homer, hid himself from the cyclops Polyphemus by clinging to the wool on the belly of a sheep. When their wicked stepmother had arranged to sacrifice the children Phrixos and Helle, a golden ram sent by Zeus swooped down from the sky, and carried them away on its back. Helle fell into the sea, but Phrixos was taken to Colchis, where he sacrificed the ram. Its golden fleece was placed upon a tree and guarded by a dragon, until it was stolen by the hero Jason, assisted by the princess Medea, and today it is used to symbolize the goal of a mystic quest.

The Hebrews were largely a nation of herders, and many patriarchs of Israel including Abraham, Moses, Jacob, and David tended flocks. Abraham was commanded by God to sacrifice his son Isaac, but was stopped at the last moment by an angel, who directed him to a ram struggling in the bushes, which he was to kill in place of the boy (Genesis 22). The story is usually interpreted as a test of faith, though some thinkers also see it as a rejection of human sacrifice. At any rate, it illustrates the close identification, almost to the point of being interchangeable, between Israel and a flock. Jewish legend tells that Moses tended flocks for 40 years, not allowing a single sheep to be hurt by wild beasts or lost. As a reward for his care, God made Moses the leader of Israel.

When Moses had placed a curse on Egypt that the first born in every home would be slain, Yahweh directed that every Hebrew household sacrifice a one-year-old male sheep or goat without blemish and smear some of the blood on the doorposts or lintel, so that Israel might be spared (Exodus 12). The Jews commemorate this event in spring during the feast of Passover, at which a lamb is eaten with unleavened bread and bitter herbs. The Last Supper of Christ, commemorated in the mass, was probably a Passover meal. In contemporary America, Easter dinner still often features roast lamb.

The one other herd animal that, though unnamed, has had a comparable role in the religious history of humanity is the scapegoat. In a passage from Leviticus, Aaron was directed to take two goats, one of which was to be sacrificed to Yahweh and the other, the scapegoat, driven into the desert for the demon Azazel (Exodus 16:7-10). This event probably reflected a residual paganism among the Hebrews, which was very promptly repudiated. A later passage in Leviticus states of the Hebrews, "They must no longer offer their sacrifices to the satyrs (i.e., goats and associated deities) in whose service they once prostituted themselves" (Exodus 17:7). The goat consecrated to Azazel has become a symbol of all that are made to suffer for the sins of the community, such as the Jews in Nazi Germany.

Both the Old and New Testaments also constantly used metaphors drawn from herding to speak of religious matters. Psalm 23, known as "The Lord's Prayer," begins

> Yahweh is my shepherd
> I lack nothing.
> In the meadows of green grass he lets me lie.
> To the waters of repose he leads me;
> There he revives my soul (1-3).

When John the Baptist first saw Jesus, he exclaimed, "Look, there is the lamb of God that takes away the sins of the world" (John 1:29). Jesus also compared God to a good shepherd (John 10), and the metaphor is commemorated in a bishop's crosier, a ceremonial shepherd's crook. In Revelations, the "Lamb" is used as a code word for Christ, who is to return for a final battle against the forces of evil. Matthew compared the Last Judgment to a shepherd separating the sheep from the goats (Matthew 25:32-33).

The growing antipathy toward goats in the Judeo-Christian tradition was in reaction to their veneration in other cultures of the ancient world. The Greek god Dionysus took the form of a goat when fleeing to Egypt to escape the serpent Typhon. A goat was sacrificed at the annual festival of

Dionysus, in a ceremony from which Greek tragic drama eventually emerged.

According to some myths, the nature spirit Pan was the offspring of the god Hermes and a goat. For other figures of the Greek pantheon, he was a sort of country bumpkin. They banished Pan from Mount Olympus for his ugliness, and so he wandered the fields and forests. When fleeing from the serpent Typhon, he tried to turn himself into a fish but was so terrified that he could not complete the transformation. He is pictured in the zodiac as a goat with the tail of a fish, the sign Capricorn, and the tale is the origin of the word "panic." Pan himself could inspire terror, the dreadful solitude of remote places, in any traveler who disturbed his mid-day sleep. The Greco-Roman satyrs, his woodland companions, had human bodies with only the ears and sometimes horns of a goat. They were often depicted pursuing nymphs, usually without much success, and they were lecherous enough to mate with animals as well.

The Greek Historian Herodotus reported that the Egyptians considered "Pan" (that is, their god Khem) the most ancient of their gods. He also claimed that in Egyptian province of Mendes, people venerated goats, especially the males, and held goatherds in great honor. To the historian's revulsion, the people of that town reportedly allowed a woman to publicly mate with a goat.

Two magical goats, which could be sacrificed, eaten, and then resurrected, accompanied the Norse god Thor. In Northern Europe, goats were admired less for their fecundity than for their ability to thrive in severe, mountainous landscapes. In the Middle Ages, however, painters often depicted the Devil with the horns of a goat.

To summarize so far, sheep and goats were increasingly used throughout the ancient world to express various polarities: sheep were generally thought of as feminine, goats as masculine; sheep were civilized, goats natural; sheep were Judeo-Christian, goats, pagan. This basic symbolism changed very little throughout the Middle Ages and the modern world, but the way in which these various qualities were valued varied greatly. In the eighteenth century, bucolic poetry often nostalgically celebrated the simple shepherd tending his flock. Among romantics of the nineteenth century, who were fascinated by the idea of primeval wildness, the goat-god Pan became by far the most popular figure in the Greco-Roman pantheon. It was a rather ironic choice, since goats actually do prodigious damage to forests by nibbling at young trees. But, as in so many other contexts from hunting to exploration, people often seem to celebrate the natural world most when engaged in destroying it.

In contrast with the West, the Chinese have never made such a great symbolic distinction between sheep and goats; in fact, the two are often interchangeable. Sheep were introduced in East Asia later than goats, and the two herd animals have often been pictured grazing together in oriental art. The eighth sign of the Chinese zodiac may be depicted as either a goat or ram. The goat, with its preference for remote solitary places, can often represent the anchorite, and the beard of a male goat resembles that often depicted on a Chinese sage. The half-legendary Huang Ch'u P'ing, who lived in the fourth century CE, was a goatherd who decided to withdraw from the world. He had meditated for forty years when his brother found him in a cave. After greeting him, the brother asked what had become of the goats. Huang Ch'u P'ing pointed to a several white stones which lay scattered around the cavern, and he began to touch them, one by one, with his staff, at which each rock jumped up and became a goat.

("Hector, the Dog," from *The Great American Christmas Book, 2007*)

9
MAN'S BEST FRIENDS

P ETS—ANIMALS KEPT IN THE HOME SOLELY TO OFFER COMPANIONSHIP to people—are largely a phenomenon of modern Western culture. Prior to that, animals were often kept in or near human dwellings, and intense bonds developed between them and people, though they almost always had either a religious or pragmatic purpose. In Rome, people would sometimes keep a snake in their home, believing it to be the spirit of an ancestor, a practice that survived into modern times in parts of Italy and Eastern Europe. Caged birds were also kept, largely to provide music. In the Middle Ages, aristocrats often kept dogs for hunting or to guard the home. Cats, as well as dogs, were used for catching mice. Hawks were used to catch rabbits and other small game. Paradoxically, as such purposes have been increasingly taken over by mechanical devices, the ownership of pets has soared. The Humane Society of the United States estimates that in 2011 39% of American households had at least one dog, while 33% had at least one cat. To some extent, these animals may serve as mediators with the natural world. Since the middle of the twentieth century, however, dogs especially have been so completely integrated into human society that it is questionable how much they can still serve that purpose. They no longer have their own detached "doghouses," but live in the home, where they are elaborately trained, fed prepared foods, and seldom allowed to run freely. Today, advertisements and popular articles often refer to people with companion animals not as "owners" but as "pet parents," while the pets themselves are called "boys" and "girls." In the late twentieth and twenty-first centuries, scholars have increasingly recognized the novelty of modern pet-keeping, and a vast academic literature has accumulated around the study of the relationships between pets and human beings. Scholars sometimes prefer the term "companion species," coined by Donna Haraway, since it embraces not only contemporary pet-keeping, but also its many precedents in the co-evolution of animals and humans.

CAT

The cat is the only animal to have succeeded in domesticating man.

—Marcel Mauss

"When I play with my cat, who knows but that she regards me more as a plaything than I do her?" wrote Michel de Montaigne in "Apology for Raymond Sebond." Touch or pet a cat and there may be sparks! Cats are constantly rubbing their backs against any available surface, so static electricity builds up in their fur. People have always been mystified by the ability of cats to survive after falling from tall trees or buildings. No wonder cats have always seemed magical. The enormous eyes of a cat shine with special intensity when the rest of its body is shrouded in darkness. Because the pupils of the cat constantly expand and contract to adjust to the level of light, they seem like the waxing and waning moon.

In many ways, cats may be subordinate to the master or mistress of the house, but their manner always suggests confidence and power. Jean Cocteau called the cat "the soul of a home made visible." The intense attachment that cats develop to their homes seems, at least in the context of most traditional cultures, feminine. The troubled partnership of cat and dog in many human homes often resembles that of women and men. We can also think of the cat within the home as the secret wildness in every person that survives despite the regimentation of our public lives. The self-assured bearing of cats suggests hidden knowledge, which people have both valued and feared.

The curvilinear design of the feline body and the cat's rhythmic way of walking are also classically feminine. Many archaic goddesses were closely associated with cats. The Greek Artemis, goddess of the moon, fled to Egypt and changed herself into a cat to escape the serpent Typhon. A panther was sacred to the goddess Astarte, the Mesopotamian equivalent of Aphrodite, who was often portrayed standing upright on her mascot. The Hindu goddess of birth, Shasti, also used a cat as her mount. Freya, the Norse goddess of love, rode in a chariot drawn by cats.

Perhaps most importantly, the Egyptian goddess Bastet was depicted with the head of a cat and the body of a woman. Our word "puss" or "pussy" for cat comes from Pasht, an alternative name for Bastet. The yearly festival of Bastet, held in autumn, was the most splendid celebration in all of Egypt. Hundreds of thousands of people would come on boats, singing and clapping to the music of castanets. They would offer sacrifices at the temple of Bastet, then feast for several days.

The Egyptians punished the unsanctioned killing of a cat with death. Diodorus Siculus reported that in the middle of the first century BCE a member of a Roman delegation to Alexandria accidentally killed a cat. A crowd stormed his house. Not even the fear of Roman anger could keep the local citizens from killing the perpetrator. Several superstitions about cats may go back to ancient Egypt, and many people still say that killing a cat brings bad luck.

According to Herodotus, the entire family in an Egyptian home would go into mourning when a cat died. All members would shave their eyebrows to show their sorrow. Dead cats were taken to the city of Bubastis, where they were embalmed and ceremoniously buried. Hundreds of thousands of mummified remains of cats have been found in Egyptian tombs.

A fable known as "The Cat Maiden," traditionally attributed to the Greek Aesop, records the triumph of feminine wiles over masculine strength, which is clear in this retelling by the British folklorist Joseph Jacobs:

> The gods and goddesses were arguing about whether it was possible
> for a thing to change its nature. "For me, nothing is impossible,"
> said Zeus, the god of thunder. "Watch, and I will prove it." With

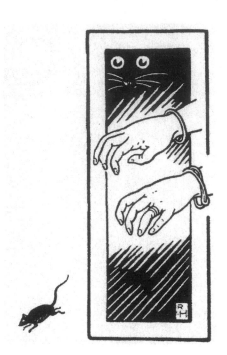

Illustration by Richard Heighway to Aesop's fable, "The Cat Maiden."

that, he picked up a mangy alley cat, changed it into a lovely young girl, had her dressed in fine clothes, instructed her in manners, and arranged for her to be married the next day. The gods and goddesses looked on invisibly at the wedding feast. "See how beautiful she is, how appropriately she behaves," said Zeus proudly. "Who could ever guess that only yesterday she was a cat!" "Just a moment," said Aphrodite, the goddess of love. With that, she let loose a mouse. The maiden immediately pounced on the mouse and began tearing it apart with her teeth.

This fable has been written down in many versions, some of which date back to the fifth century BCE in Greece.

"Dick Whittington and His Cat," a tale from Medieval England, shows how cats were valued in the early modern period by those engaged in trade. The hero, Dick Whittington, was an impoverished young man in London who had worked hard and managed to buy a cat, which he lent to a ship's captain. The captain sold the cat for a vast fortune to the king of the Moors, whose palace was plagued by rats. Whittington became a wealthy man and, eventually, Lord Mayor of London in the late thirteenth and early fourteenth centuries, though the tale was not written down until much later.

Aboard ships, cats were kept as mascots and served to catch mice. Virtually all mariners were male, and they sometimes believed that the presence of a woman on board, or even the mention of a woman's name, would bring ill luck. The cat, often the only female on the ship, was a mediator with the feminine powers of the weather and the sea. Mariners predicted the weather by watching the cat. When the cat washed its face, they would expect rain. When the cat was frisky, they would expect strong winds. Cats would also know if the ship was about to sink. Every detail of the cat's behavior would be closely scrutinized for portents.

Superstitions about cats are almost as diverse as they are numerous. A black cat, for example, is usually thought of as a sign of bad luck, while a white cat means good luck. Sometimes, however, especially in England, this has been reversed. Wives of mariners in England would keep black cats as a charm for the safe return of their husbands at sea, a practice that people in other communities could misinterpret as witchcraft.

In Renaissance Europe, cats were often thought to be the familiars of witches, and black cats in particular were frequently named as such in the witch trials. Jean Boille, who was burned as a sorceress at Vesoul in 1620, claimed to have seen demons and cats participating together in sexual orgies at the witches' Sabbath. A pact with the Devil was sealed with a paw print placed on the body of a witch. The Black Witch of Fraddan flew through the air at night on an enormous cat. In the early thirteenth century, the bishop of Paris, Guillaume d'Auvergne, claimed that Satan appeared to his followers in the form of a black cat and they had to kiss him beneath the tail.

Diabolic, and sometimes almost as frightening as the Devil himself, is the King of the Cats in Irish folklore. Sometimes the King is black and wears a silver chain, but he cannot always be recognized. Lady Wilde in *Legends of Ancient Ireland* tells of a man who once, in a fit of temper, cut off the head of a domestic cat and threw it into a fire. The eyes of the cat continued to glare at him from within the flames, and the feline voice swore revenge. A short time later the man was playing with a pet kitten; suddenly the kitten lunged, bit his throat, and killed him.

When people are fond of certain animals, they assume the animals will also be beloved by the gods and goddesses, and they offer them up as sacrifices. The ancient Egyptians may have punished the killing of a cat outside of a temple with death, but they offered thousands of cats to Bastet, generally by breaking their necks. Christianity officially rejected animal sacrifice, but ritual killing of cats continued for thousands of years. Cats were burned alive on Ash Wednesday in Metz and other Continental cities during the Middle Ages to produce ash needed for the mass. In England, the effigy of Guy Fawkes that was ceremonially burned every year sometimes contained a cat that would howl as the flames rose. Remains of cats that were walled up alive have been found in the foundations of several Medieval buildings, including the Tower of London. This was the theme of Edgar Allan Poe's famous horror story "The Black Cat." Terrified that his wife was a witch, and that her black cat was the Devil, the narrator killed his wife and built a wall to conceal her body, but the cat howled from behind the wall until the police came.

Toward the end of the Middle Ages, there were few cats left in Europe. Their absence led to a great increase in rats and murine diseases, most notably bubonic plague. The few cats that had survived the persecutions came to be highly valued. Europeans realized that cats were not only useful but also loyal and affectionate, and thus benevolent cats began to appear in fairy tales, though they still usually seemed to have something a little disturbing about them. In "The White Cat" by Madame D'Aulnoy, a magical feline guided the hero through all sorts of trials and tribulations. Finally, the cat cast off its skin, became a woman, and married him. She then burned the skin; after all, would the man really want his wife changing shape and casting spells? Perhaps the magic here is the power of young love, a kind of sorcery that needs to put away as a person enters maturity.

In "Puss in Boots" by Charles Perrault, a cat loyally helped a young man. To win a fortune for him, however, the two had to connive and deceive everybody else. Master Puss made up a title, "the Marquis of Carrabas," for the young man. Then the cat told harvesters that they would be chopped up into little pieces if they didn't tell the king their land belonged to that Marquis. When challenged by the cat to display his prowess in shape-shifting, the ogre, who really owned the land in question, transformed himself into a mouse. The cat immediately pounced on the mouse, ate him, and took over the ogre's castle for the young man. Finally, the young man had so much wealth that he could marry the king's daughter. If the story were told from another point of view—say, that of the ogre—the reader could easily take this cat for the Devil. Still, it is great having such a cat on your side!

In folklore, the animals in a household often make up their own little society, a sort of microcosm. The dog, of course, is among the most domesticated of animals, while the rodents are completely wild. The cat is in between. The folkloric dog and cat are constantly quarreling and making up. Sometimes they cooperate to help their master, but the old enmity can break out at any time. The cat and mouse, by contrast, are mortal enemies. The mice in the household hardly ever defeat the cat, though they often manage to get away. The situation is a bit like a troubled family of human beings, where mother and father quarrel, and the children suffer.

Buddhists generally dislike the cat, though they have seldom carried this to the extremes we find in the West. The Jatakas, ancient Buddhist fables describing how the animals assembled around the deathbed of Buddha to pay him homage, note that the cat was taking a nap and didn't come. According to another traditional tale, the goddess Maya sent a rat with medicine for the ailing Buddha, but the cat killed the rat, so Buddha perished. Nevertheless, cats were regularly kept as mousers in households of China, Japan, and other countries of the Far East. Artists were often fascinated by their alertness, as well as by their sensitivity to subtle sounds and motions. For such a common animal, cats were notably absent from the Chinese zodiac, in part because they were closely associated with the element of earth.

For all their differences, Christianity and Buddhism have both tended to be suspicious of archaic magic. Perhaps this is partly the reason the cultures that have grown around these religions so often view the cat, the most magical of animals, with mistrust. Islam may be a legalistic religion, yet the faith delights in extravagant tales of the supernatural; consequently, Muslims have always been very fond of cats. In one legend, Muhammad once found his cat Meuzza sleeping on his robe. So as not to disturb his pet, the prophet cut off a sleeve and put on the rest of the garment. When he returned, Meuzza bowed to him in gratitude. Mohammed blessed the cat and her descendants with the ability to fall and land on their feet. When cats enter a mosque, it means good luck for the community. In one story from Oman, told by Inea Bushnaq, a cat caught a mouse and was about to devour it; the mouse begged to be allowed a prayer before death. When the cat agreed, the mouse suggested that the cat pray as well. The cat raised its arms in prayer, and the rat escaped. When a cat rubs its face, the story concluded, it is remembering the smell of the rat.

At the beginning of the nineteenth century, the German writer E. T. A. Hoffmann took on the formidable task of trying to imagine the feelings of a cat in *The Life and Opinions of Kater Murr* (1820). A passionate if somewhat reluctant romantic, Hoffmann felt cats were like those who work magic in verse or paint. Like artists, cats have mysterious insights. Like artists, cats often seem vain and impractical. Both cats and artists have an odd combination of innocence and guile. The tomcat Murr, who tells his story, affectionately mocks his master. He has adventures climbing the rooftops of the town. He reminds the reader in his preface, "Should anybody be bold enough to raise doubts concerning the worth of this extraordinary book, he should consider that he confronts a tomcat with spirit, understanding, and sharp claws."

Poets always love mystery, and so they also love cats. W. B. Yeats and T. S. Eliot are among the many who have found inspiration in cats, but the most famous poem of all about cats is "My Cat Jeoffry" by Christopher Smart. The author takes precisely the characteristics that have impressed people as diabolic and uses them to make the cat a symbol of Christ:

> For he keeps the Lord's watch in the night against the adversary.
> For he counteracts the powers of darkness by his electrical skin &
> glaring eyes.
> For he counteracts the Devil, who is death, by brisking about the
> life.

For Smart, the many paradoxes that surround the cat are proof of divinity.

In the decades immediately following World War II, people in the United States and Europe liked to romanticize alienation. In the slang of the Beatnik movement, a "cat" became somebody who preferred the colorful life of the streets to the mainstream of American society. In the last few decades of the twentieth century, cats temporarily replaced dogs as the most popular pet in the United States. Some reasons for this preference are pragmatic. Cats are smaller, eat less, need less space to exercise, demand less human attention, and are less expensive to care for than dogs. For those who find the emotional exuberance of dogs embarrassing, cats seem to offer emotional support without sacrifice of decorum. The relationship of cats to people can be warm and nurturing yet with a distance of respect; intimate yet full of riddles.

Nevertheless, cats kill billions of small birds and mammals every year in Europe and North America. The philosopher Jacques Derrida had an epiphany when his cat came upon him naked and he felt ashamed. This inspired an essay, *The Animal that I therefore Am*, in which he explained that, while his own existence might not be assured, his cat's was beyond question. Perhaps Derrida's shame was a refined sort of fear, and his experience reflected the ancestral memory of a time when cats were mighty predators and human beings were prey.

DOG

> *Ay, in the catalogue ye go for men;*
> *As hounds, and greyhounds, mongrels, spaniels, curs,*
> *Shoughs, water-rungs, and demi-wolves, are clept*
> *All by the name of dogs. . . .*
>
> —WILLIAM SHAKESPEARE, *Macbeth*

Around 12,000 BCE in Eurasia—though according to some ethologists, much earlier—the dog became the first animal to be domesticated by human beings. Cats continue to appear wild even when raised in the family living room. Sheep and cattle generally stay together in herds, even under human direction. In the continual war between man and nature, only dogs appear to be unequivocally on our side. According to a legend of the Tehuelche Indians, after the sun god had created the first man and woman, the deity immediately created a dog to keep them company. Emotionally, dogs seem akin to human beings. Some people believe that dogs are the only animals apart from humans that can feel guilt. Others dismiss that perception as an anthropomorphic illusion or even as canine hypocrisy. People often regard dogs as icons of either the faithful companion or the sycophant. The dog joins the realms of culture and nature; in much the same way, the mythic dog serves as a mediator between life and death.

Already in ancient Egypt, dogs and cats were the most beloved of pets. According to Herodotus, when the family dog died, all members of the household would shave their whole bodies, including their heads, to show their mourning. Many Egyptian pictures have come down to us of people caressing dogs, as well as using them in the hunt. While cats were associated with both the sun god Ra and the goddess Bastet, dogs were associated with the underworld and with death. The appearance of the Dog Star, Sirius, was a sign that people should prepare for the rising of the Nile River. (The association of dogs with the star Sirius reaches all the way from Egypt to Mexico and China.) Plutarch, however, reported in his essay "Isis and Osiris" that when the blasphemous conqueror from Persia, Cambyses, had slain the sacred bull Apis, only dogs would eat the body, and so the dog lost its status as the most honored animal among Egyptians.

Tombs in which owners were interred with their dogs, either as canine effigies or in the flesh, have been found throughout the ancient world. They are not only common throughout Eurasia and parts of Africa but also in pre-Columbian America. Just as dogs led hunters tracking game through the wilderness, they were expected to guide people through the next world.

Their keen sense of smell gave dogs an ability to guide people in the hunt. A dog could find the location of game that was not even remotely visible. After the hunt, a dog guided people through the woods back to their settlement. We should remember that this was long before the use of the compass or of even remotely accurate maps. This ability must have impressed

HIS MASTER'S VOICE

This ad from the early twentieth century exploits the dog's reputation for fidelity.

(Courtesty of RCA)

people as miraculous. Small wonder that a vast range of cultures on every continent have regarded dogs as guides to the world after death!

Though dogs are occasionally seen as solar animals, they are usually associated, like cats, with the moon. Perhaps this is because they howl at the moon, as do their relatives—wolves, coyotes, and jackals. Many cultures understand the howling of dogs as an omen of death. According to Jewish tradition, they can see the angel of death. In Greek mythology, they are companions of the lunar goddesses Artemis and Hecate. In Virgil's *Aeneid*, dogs howl at the approach of the darker and more magical lunar goddess Hecate, who herself is accompanied by dogs. Several traditions also make dogs the guardians of the underworld. The best known is Cerberus, who keeps watch at the entrance to Hades in Greco-Roman mythology. According to Hesiod, this dog had fifty heads, though later writers reduced the number to three.

In Norse mythology, the abode of the dead is watched over by the enormous dog Garm. When the final battle at the end of the world comes, Garm will swallow the moon. This monstrous dog will finally do battle with the Tyr, the god of battles, and both will be slain. In Hinduism and Buddhism, two dogs accompany Yama, the lord of the dead. They each have four eyes, and serve their master by searching out those who are about to die. In Aztec mythology, the departed soul descended to the underworld, and then came to a river guarded by a yellow dog. In European folklore, demonic dogs accompanied the Wild Huntsman across the sky in his search for lost souls. To even hear the hounds meant that you would die soon. In the lore of West England, the Devil's Dandy Dogs passed over the moors during storms, breathing fire and tearing hapless strangers to pieces.

In the religion of the Aztecs, the canine diety Xolotl was also intimately associated with the world of the dead. According to an Aztec legend, human beings had once died out completely, and the gods wished to bring them back. Xolotl traveled beneath the ground to obtain the bones of the departed people. This intrusion angered the god of the dead, who pursued Xolotl, making him stumble, fall, and break the bones in many pieces. Xolotl recovered the bones and brought them back to surface of the Earth. When the gods sprinkled the bones with their own blood, the pieces became living men and women of many shapes and sizes.

A further reason for the association of dogs with death is that packs of feral dogs roamed the entire ancient world, searching for carrion, which included the bodies of human beings. Perhaps the greatest disgrace for an individual in almost cultures of the Mediterranean was to have his/her corpse eaten by dogs. In Homer's *The Illiad*, the Trojans dreaded that this would be the fate of Hector's body, after their champion was killed by Achilles. In Sophocles' *Antigone*, the heroine feared that would happen to her brother, Polynices, if he were not given a proper burial. In Ovid's *Metamorphosis*, the hunter Acteon experienced an especially demeaning death, changed by the goddess Diana into a stag and killed by his own hounds. Because Jezebel, wife of King Ahab of Israel, spread the worship of Baal, the prophet Elijah prophesied that dogs would devour her body. Jehu later ordered her thrown down from a window, and those who went to bury her found only the skull, feet, and hands.

Lady Wilde has written of dogs in Ireland: "The peasants believe that the domestic animals know all about us, especially the dog and the cat. They listen to everything that is said; they watch the expression of the face and can even read the thoughts. The Irish say it is not safe to ask a question of a dog, for he may answer, and should he do so the questioner will surely die." The dog certainly shares the life of human society more intimately than any other animal. This, of itself, can make people feel uneasy. Human beings view dogs with a strange combination of affection and contempt, of domination and fear.

The name of Cuchulainn, the popular hero of Celtic myth, literally means, "hound of Culann." When he killed a smith's ferocious hound, Cuchulainn had to take on himself the role of the creature he had killed. When roused to battle, his appearance changed. His eyes bulged or contracted. His jaw opened from ear to ear, like that of a dog, while a light like the moon rose in his head. When three witches in the form of crows tricked him into eating the flesh of a dog, as well as violating other taboos, Cuchulainn was soon killed.

Dogs were held in special reverence in Persia. According to legend, Cyrus, who founded the Persian Empire, was left out to die at birth, but then saved by a bitch that suckled him. In the religion of Zoroaster, which began in Persia, a dog had to accompany a funeral procession to ensure that the deceased had a peaceful journey to the next world. The Zoroastrians believed that dogs were able to see spirits, and so they could protect families from evil powers beyond the awareness of human beings. In gratitude for this defense, families were expected to feed hungry dogs, using ritualistically prepared food. The members of the family then said prayers as the dog ate. Dogs guarded the Cinvat Bridge that led to the next world, protecting the righteous but leaving the unrighteous to demons. In the religion of Mithras, the major rival to Christianity in the latter Roman Empire, a dog would be among the animals to accompany Mithras during the sacrifice of a great bull to rejuvenate the world. After the sacrifice, the dog would lap up the blood that had been spilled.

Just as the dog guards the home, people in the ancient world also thought of dogs as guarding the body from demons or disease. In Mesopotamia, the dog was sacred to Gula, the Babylonian goddess of healing. At times, she was represented as a bitch suckling her pups. In her human form, dogs accompanied her. Many dog figurines have been found in her temple, and they were used to ward off illness. In Greece, a dog generally accompanied Asclepius, the legendary doctor who once raised a man from the dead.

Dogs were, for the most part, favorably regarded in the Greco-Roman world. Many people find the most touching scene in Homer's *The Odyssey* occurred when the hero finally returned home and was recognized only by his hound, Argos. The dog wagged his tail and then died. The Greeks and Romans sometimes wrote very affectionate epitaphs for their dogs.

The philosopher Diogenes, a contemporary of Alexander the Great, called himself a "hound." Members of his school were known as "cynics," after the Greek word for "doglike." Like dogs, they lived in the society yet did not fully belong to it. Since then, dogs have often symbolized alienation. Diogenes not only praised the fidelity and the modest needs of dogs but also their lack of shame, since they would not hesitate to urinate or copulate in public.

Ancient authors such as Ctesias and Pliny the Elder wrote of the cyno-cephali, who had human bodies and the heads of dogs. These figures ulti-mately go back to the jackal-headed Egyptian deity Anubis. After Alexander had conquered Egypt, his soldiers conflated Anubis with their God Hermes, since both were guides for the dead. The soldiers called this composite deity Hermanubis, and erected a temple to him in Alexandria, which became one of the most popular shrines in the ancient world. In time, his cult was finally absorbed into Christianity, and Hermanubis became Saint Christopher, who is often portrayed with the head of a dog.

In the Middle Ages, Saint Roche, who is invoked against diseases, was depicted with a dog. This holy man worked with victims of the bubonic plague. One day, however, he himself was with stricken and sores appeared on his body. Saint Roche wandered into the woods to die, when a dog came up and licked the infection. With the help of the dog, which also brought him bread, Saint Roche miraculously recovered.

In Medieval burials, there is a remnant of the tradition in which the dog serves as a guardian to the next world. The lord and lady of the house would often be buried with their dogs. Splendid sculptures and brass reliefs on graves show the deceased stretched out with a faithful dog at his or her feet. Today, dogs are often buried in pet cemeteries, seldom with their masters, yet many people still hope to be reunited with a beloved pet in the world beyond.

Throughout the Middle Ages, travelers spread accounts of dog-men in distant lands, especially the mythical kingdom of Prester John in India. One popular legend is that Saint Christopher was from a race of savage cynocephali, who ate human flesh. They could not speak but only bark. However, in answer to his wordless prayers, God granted Christopher human speech. Another tale accounts for his strange appearance by telling that Saint Christopher was once extraordinarily handsome, but prayed to God for the head of a dog, so that women would leave him in peace.

A version of this deity also entered Chinese legend. One very popular tale tells how a dog married a princess. Barbarians had invaded from the west, and the desperate Emperor promised that anybody who could drive back the enemy might marry his daughter. A dog heard the pledge, crept behind enemy lines and killed the opposing commander; he chewed off his victim's head, brought it back, and presented it to the Emperor. When they discovered what had happened, the barbarians withdrew. The dog, which could speak like a human being, then reminded the Emperor of his promise. When the Emperor objected that marriage between a person and an animal was impossible, the dog replied that he could be made human by being placed under a bell for 280 days, provided that nobody disturbed him in the interim. This was done, but when only one day remained the Emperor was overcome with curiosity and lifted the bell, to see a creature with a human body and a canine head. The marriage went ahead as planned, and the tribe known as the Fong of Fuzhou claim to be descended from the couple.

Closeness to humanity has by no means necessarily worked to the advantage of dogs. We often try to judge dogs by human standards, which may not always be appropriate. We include them in human hierarchies, where they are at or near the bottom of the scale. The epithet "dog" traditionally suggests a combination of contempt and mistrust, such as masters would feel for their slaves. We use the term "ass-kisser," taken from the greeting behavior of dogs, to describe hypocritically servile people. "Bitch" is used as a derogatory term for an overbearing woman or an effeminate male.

Partly in reaction to other cultures, especially that of Egypt, the Hebrews felt a repugnance for the dog. Not only is the dog an "unclean" animal in the Old Testament, but a revulsion against the dog is expressed repeatedly in very graphic terms: "As a dog returns to its vomit, so a fool reverts to his folly" (Proverbs 26:11). The view in the New Testament is not much more generous. Revelations lists "dogs" (which might be a metaphor) among those who must remain outside the kingdom of Heaven, together with "fortunetellers," "fornicators," "murderers," "idolaters," and "everyone of false speech and false life" (22:15).

Dogs are often compared to the enemies of Israel in the Old Testament:

> Yahweh, God of Sabaoth, God of Israel,
> Up, now, and punish these pagans, show no mercy to
> these villains and traitors!
> Back they come at nightfall,
> Snarling like curs,
> Prowling through the town (Psalms 59:5-6).

The Hebrews, who were very fastidious about the preparation of food, insisted that animals be slaughtered according to prescribed rituals. In addition to taking a very dim view of the hunt in general, the people of Israel regarded meat touched by hunting dogs as unclean.

Islam also takes a negative view of dogs, though there are noteworthy exceptions. Muslem tradition places nine animals in heaven, including two dogs. One is the dog of the apocryphal prophet Tobit. The other is Kasmir, the dog of the Seven Sleepers of Ephesus, from a Christian legend that passed over into Islam. Seven young Christians took refuge in a cave to escape persecution by the Roman soldiers during the reign of the Emperor Decius. They slept for two hundred years. After waking, one of them went into town to purchase provisions. He was amazed to find that almost everyone had converted to Christianity. According to the Koran, Kasmir kept watch outside the cave for the entire time, neither eating, drinking, nor sleeping.

Even in Europe, it took some time for dogs to be fully accepted as pets, especially among the middle and lower classes. In 1613, Margaret Barclay of Scotland was tried for witchcraft. With the assistance of another woman, Isobel Insh, and in the company of a black lap dog, she had allegedly made a clay image of mariners and their boat one night. Then, together with the dog, she had gone down to the shore and cast these images into the waves. Immediately, the water had turned red and the sea had begun to rage. At about the same time, a ship had gone down near the coast, killing all the crew except two men. The daughter of Isabel Insh, a girl of only eight, was called in to testify, and claimed to have witnessed the witchcraft, adding that her mother been present only at the making of the clay images but not when the spell was cast. The child added that the dog gave off fire from his jaws and mouth to illuminate the scene. Margaret Barclay was forced to confess under torture. Though she later retracted the confession, she was executed.

But absolute fidelity to one's lord and master, exemplified in faithful dogs, was an especially central virtue of the ancient and feudal worlds. There are countless stories throughout the world of dogs that kill themselves after their masters have died, often by refusing all food. Pliny the Elder wrote of a dog named Hyrcanus that threw itself on the blazing funeral pyre of his master. The most famous of all is Hachiko, an akita dog that would go to meet his owner, Professor Ueno, every day after work at the Shibuya train station near Tokyo for about a year. One spring day in 1925, the professor did not return, for he had unexpectedly died. Until its own death almost ten years later, Hachiko continued to return to the station to wait. There is now a monument to Hachiko in the Shibuya train station, and this dog is upheld as a model of family loyalty to Japanese children.

Seldom do people pass the test of canine loyalty. One of the few who did is Yudhishthira, in Hindu myth, when he ascended to heaven without his dog. There was thunder then a great light. He could see the god Indira waiting for him in the divine chariot. Invited to enter, Yudhisthira stepped aside so the dog might go first. Indira objected, saying that the presence of a dog would defile heaven. Yudhisthira replied that he could conceive no greater crime than to send the faithful dog away. At that moment, the dog was transformed into Dharma, the god of righteousness. The words of Indira had been a final test, and Yudhisthira had shown his worthiness through fidelity to his companion.

Just as many tales celebrate the fidelity of dogs, others lament the inability of human beings to reciprocate this loyalty. A good example is the tale of Guinefort, a greyhound on the estate of Villars near Lyons in France. A knight had left his baby to be guarded by Guinefort, and then later returned to find the nursery a bloody mess. The baby was missing, and Guinefort greeted the man with bloody jaws. The outraged father killed Guinefort, but then soon heard the child crying. On finding the mangled body of a snake, he realized that Guinefort had actually saved the infant from an attacking viper. The grave of the dog became a site of pilgrimages, where parents would bring sickly or deformed children to be healed. Monks in a nearby monastery looked on in consternation while peasant women prayed to the dog, hung swaddling clothes in nearby bushes, and practiced what seemed to be pagan rituals.

The treatment of dogs not only reflects, but often exaggerates, the social norms of their society. In aristocratic societies, human beings were valued largely according to their ancestry. Like kings, queens, and nobles, the thoroughbred dogs in noble houses had recorded bloodlines that went back several generations and fabricated ones that went back to remote antiquity, often to King Solomon or even Adam. A bit ironically, the pedigrees of dogs gained in importance with the rise of the middle class in the eighteenth through twentieth centuries, precisely as aristocratic ties of blood lost their significance.

With the rise of the middle class in Victorian times, the unconditional fidelity of dogs became a nostalgic reminder of the Middle Ages, a period that was widely idealized. Since one could no longer demand such loyalty of people, one valued the virtue all the more in hounds. One story that was constantly retold is that of the "Dog of Montargis," which belonged to Aubry de Montdidier, a favorite courtier of the French King Charles V. The dog's master was murdered in the wood of Montargis near Orleans in 1371.

The dog was the only witness and followed the murderer, Robert Macaire, everywhere, constantly barking at him in an accusatory manner. Finally, a duel between the dog and man was arranged. After being badly defeated, Macaire confessed to his crime and was executed.

The Nazis wanted to promote unquestioning loyalty and obedience, a quality that found best exemplified in dogs. Hitler once said that he trusted nobody but his girlfriend, Eva, and his dog, Blondie. Then, when society had thoroughly industrialized, dogs served as nostalgic reminders of the rural past. Especially in the 1950s and early 1960s, such dogs such as Lassie and Rin Tin Tin were enormously popular on television.

Our taboo against cannibalism extends to dogs, but only to a limited extent. Until very recent times, dogs were eaten almost everywhere, regularly by the poor and by far more people when food became scarce. There were dog slaughter houses, and dog meat was available at some delicatessens and restaurants. Germans, for example, ate about 84 tons of dog meat or more every year in the decade leading to World War I; that figure increased by almost a third during the Great War, briefly declined, and then shot up even more during the Great Depression of the 1930s. The phrase "dog eat dog," referring to ruthless, unrestrained competition, originated in the United States during the Great Depression, and probably originally referred to this practice.

The Dog Catcher remained a fixture of everyday life in Western countries at least until the second half of the twentieth century, and feral dogs that were captured might at times be not only killed but also eaten. Only since the end of the World War II, the consumption of dogs throughout Europe and North America has become miniscule, and the professions of Dog Butcher and Dog Catcher have almost disappeared (though we still have Animal Control Officers, with far broader responsibilities). In North Africa and East Asia, people continue to eat dogs, but Westerners usually find this utterly uncivilized, and governments usually try to discourage it.

Feral dogs have also become relatively rare in the West, though workers in animal shelters continue to agonize over whether, and when, dogs should be euthanized. Packs of such dogs remain common in much of the world, and, in villages throughout much of Africa, people who adopt a dog do so not by taking it into their homes but simply by feeding and befriending it.

In early space exploration, dogs were often used as surrogates for human beings. The first living creature to be sent into orbit was a Samoyed named Laika, sent up in a Soviet satellite in 1957. After six days, the oxygen ran out and she died, but her corpse continues in orbit to this day. The death evoked wide protests, and Laika has come to symbolize the fragility of all life in this age of scientific exploration. In the next few decades, however, computers, which had attained great sophistication, largely replaced dogs, other animals, and, to an extent, even people in outer space.

As the pragmatic usefulness of the dog in our daily lives is gradually reduced, the symbolic importance of dogs may even be increasing. In this age of electronic security systems, dogs are relatively inefficient at guarding the home. Nevertheless, most people have seen the cartoon dog McGruff, in a trench coat and floppy hat, who tells people on television to "take a bite out of crime." Dogs are used to sell a vast variety of products related to security, from alarms to programs against computer viruses.

Today, dogs are almost completely integrated into the consumer culture of Europe and North America. Dogs now have special gyms, fashions, gourmet foods, beauty parlors, jewelry, five star hotels, television shows, therapists, and almost everything else that people have. But dogs, like us, pay for all of this luxury with freedom. In the last decades of the twentieth century, most urban communities have prohibited dogs from running free even in city parks.

Not everybody likes dogs, but those who do are very passionate about them. Dogs often appear helpless, yet they are usually able to take care of themselves pretty well. This combination of vulnerability and strength makes dogs, for good or ill, so very "human."

10
BEASTS OF BURDEN

T IS IMPOSSIBLE TO USE ANIMALS WITHOUT "HUMANIZING" THEM, and beasts of burden may be the best example of this. It is generally easy for human beings, especially manual laborers, to identify with animals such as horses, mules, or llamas. Their burdens are a metaphor for the challenges we face constantly in endeavoring to survive and prosper. We speak of "financial burdens" and "emotional burdens," as well as physical ones. From a human point of view, these creatures are generally accorded a status that corresponds to what they carry. Those that bear human beings, such as horses, generally rank higher than those that, like mules, transport luggage. The legendary mounts of heroes—such as Al-Borak which carried Mohammad to Heaven—are, of course, the most prestigious of all.

ASS, CAMEL LLAMA, AND MULE

> *Orientis partibus*
> *Adventavit Asinus,*
> *Pulcher et fortissimus.*
> *Sarcinis aptissimus.*
> > *(From the East,*
> > *the ass approached,*
> > *lovely and very strong,*
> > *the best baggage carrier)*
>
> —Carol sung at the Medieval Feast of the Ass at Beauvais, France

The donkey has a reputation as a sort of holy fool, strangely poised between wisdom and stupidity. A charlatan exhibits a supposedly learned donkey at a fair.

(Illustration by
J. J. Grandsville, from
Fables de la Fontaine,
1839)

The ass and camel are, for the most part, peaceful animals that help with daily tasks, while the horse excels in arts of war. The ass and camel both have greater endurance than the horse, though they are not as fast. The camel thrives especially well in hot, dry climates, and the ass is very sure-footed in mountains. The ancient Mesopotamians noticed that the cross between a mare with a jackass would produce a mule, which had many advantages of both, though it has sometimes been stigmatized as a product of an "unnatural" union.

The *Avesta*, a scripture of the Zoroastrians, told of an ass with three legs, six eyes, nine mouths, and a single horn. This animal was as large as a mountain and stood in the middle of a wild sea, whose waters it forever purified. This early unicorn symbolizes the primeval innocence of a time when the world was new, but it also shows the awe with which the ass was once regarded.

The ass, or donkey, was first domesticated in ancient Egypt around 3000 BCE, well over a millennium before the horse. In the Bible, God enjoins Job to marvel at the difference between his donkeys, those in the wild, and those kept by human beings:

Who gave the wild donkey his freedom,
>and untied the rope from his proud neck?

I have given him the desert as a home,
>the salt plains as his own habitat.

He scorns the turmoil of the town:
>there are no shouts from a driver for him to listen for.

In the mountains are the pastures that he ranges
>in quest of any green blade or leaf (Job 29: 5-8).

The ass survives precariously in the wild today, but very few people think of that animal as anything but tame. It began to acquire new associations through domestication, without casting off the old ones, until it became one of the most complex animal symbols of all.

The Hebrews had a special affection for the ass because of the unstinting service it performed, much as the Arabs did for the camel. According to the classifications in Leviticus, the ass was an "unclean" animal, yet, in contrast to the pig, the people of Israel never regarded it with revulsion. After the asses carried the Israelites and their possessions from slavery in Egypt, Yahweh ordered that the first born of every ass be consecrated with the sacrifice of a lamb.

Once the King of Moab summoned the magician Baalam to place a curse on the Israelites. Baalam set out on his she-donkey, when an angel, sword in hand, appeared in his path. The donkey turned aside from the road, and Baalam beat her to draw her back. This happened a second time, and then a third. At this, Baalam picked up a stick and began to strike the donkey furiously. The donkey reproached Baalam saying, "Have I not carried you since you were a young man? Have I ever failed you? Why do you beat me now?" Then Baalam looked up and saw the angel. "It is lucky for you," the angel told him, "that your donkey saw me, though you did not, and turned aside. Had you continued, I would have killed you, but I would have let the donkey live." This story, from the Old Testament (Numbers 22:22-35), is one of the earliest and most explicit condemnations of cruelty

Twelfth-century French depiction of a troubadour as a donkey. It is very similar to illustrations from the ancient Near East.

to animals in the ancient world. For us, perhaps there is another lesson: If people knew a bit more about the proud history of the ass in human culture, perhaps being called an "ass" would be taken as a compliment rather than an insult.

The ass was used for work in vineyards and was sacred to Dionysus, the Greek god of wine. The Greeks, however, generally associated the ass with the Phrygians, their traditional enemies. In one legend, King Midas, a follower of Dionysus and king of Phrygia, failed to appreciate the music of Apollo. "You have the ears of an ass," said the god of the sun to Midas. When Midas looked into the water, he saw that his ears had grown long and hairy. The ears of a donkey became a familiar symbol of stupidity. The "fool's cap" used by jesters in Medieval Europe had two points with bells, symbolizing of the ears of a donkey. In Carol Collodi's *Pinocchio* (1882), the hero, a naughty marionette who wants to become a real boy, is close to having his wish, when he decides to play rather than go to school. Like other lazy children, he soon grows donkey ears. After a while, he is transformed entirely, not into a boy but into a donkey, and must go through several further adventures before finally becoming a human being.

In the fables of Aesop, the ass is always a loser. In one tale, an ass put on the skin of a lion and roamed about frightening man and beast. A fox heard him braying and said, "You would have scared me too . . . if I had not heard your braying." Socrates, in Plato's dialogue "Phaedo," stated that a person who is too concerned with bodily pains or pleasures might, after death, be reincarnated as a donkey.

The novel *The Golden Ass* by Lucius Apuleius, written in Rome during the first century CE, exploited, but also looked beyond, the image of the donkey as a fool. The hero, Lucius, had an affair with the maid of a great sorceress. When the couple saw the mistress of the house turn herself into an owl and fly away at night, Lucius wanted to do the same. His mistress gave him a potion that was supposed to turn him into a bird. It was the wrong charm, and Lucius became a donkey. His mistress told him that he could only be turned back into his true form by eating roses. When he tried to nibble some roses at an altar, the stable boy chased him away. Thus began a long series of misadventures, in which Lucius as a donkey was beaten, forced to carry heavy sacks, and even sexually abused. As a beast, he learned humility and wisdom. When his ordeal was completed, he received roses from a priest of the goddess Isis, turned back into a man, and was initiated into Egyptian mysteries. One interpretation of the story is that the form of an ass represents the physical body, imprisoning yet challenging the human spirit.

In popular Christian iconography, an ass, together with an ox, attends the infant Jesus in the manger. This is actually not mentioned in the Bible, and the image is probably inspired by the words of the prophet Isaiah that, "The ox knows its owner and the ass its master's crib" (Isaiah 1:3). An ass also carries the Holy Family to safety in Egypt, and one later bears Jesus on his entry into Jerusalem. When the ass lost prestige, Christians often understood Jesus' the choice of a mount as a sign of his humility. For the contemporaries of Jesus, however, riding an ass was a sign of majesty, since it was still the mount of kings. Well into the Middle Ages and the Renaissance, clergy preferred to ride on an ass rather than a horse in emulation of Christ the King. Legend has it that the ass still bears a dark cross over its shoulders as a symbol of Christ's passion.

The ass appealed to the Medieval love of paradoxes. They understood that what appeared to be foolishness could sometimes be holy innocence, even wisdom. A pageant known as the "Festival of the Ass" became popular in the northwestern parts of Medieval Europe. In Beauvais, France, a splendidly dressed maiden, representing Mary and bearing an image of Christ, was seated on an ass, for a magnificent procession from the cathedral to the parish church of Saint Stephen. Instead of praises of Christ, however, the choir sang a hymn to the ass. At the end, instead of saying *Deo gratias* or "Thanks to God," the congregation would say "hee-haw, hee-haw, hee-haw." The clergy, of course, often complained about the apparently sacrilegious ceremonies. Nevertheless, the festival was tolerated, in the name of both religious tradition and fun.

For many animals in Western tradition, popular symbolism has combined radically different traditions of Christianity and paganism, and the ass may be the best example of all. The pagan tradition, which made the ass an object of mockery, has been more dominant. But there was usually affection behind the mockery. The ass, unlike the horse, was very rarely accused of being the familiar of a witch or tried in Medieval courts.

In ancient Mesopotamian art, we often see the motif of a donkey standing upright on his hind legs, playing the harp, and singing. This motif was taken back to Europe by crusaders during the Middle Ages, where it symbolized the divine folly of love. Shakespeare uses this motif in his play *A Midsummer Night's Dream*. The mischievous fairy Puck gave Bottom, a simple tradesman, the head of an ass. Bewitched by a magic potion, the fairy queen Titania doted on Bottom for an evening; after the charm wore off, she was abashed and no longer dared to defy her husband.

A worldly equivalent is the donkey whose excrement was gold, which appeared in the "Donkey Skin" by Charles Perrault and many other European fairy tales. The donkey again combined something wonderful with something ridiculous. In the fairy tale "The Magic Table, the Gold Donkey and the Club in the Sack" by the Grimm brothers, the image was, so to speak, "cleaned up." A donkey spewed gold from its mouth whenever a person said the word "Bricklebrit!" The image was a strange anticipation of the "automatic teller" machine used today.

The tradesman Bottom with the fairy Titania in Shakespeare's "A Midsummer Night's Dream."

(Illustration by Arthur Rackham)

The lore of the ass is full of wonders, in East as well as West, yet these have almost never been without a touch of humor. The Taoist immortal Chang Kwo-lao is an elderly gentleman who rides a donkey for vast distances every day. Whenever a journey is finished, he folds the donkey up like a piece of paper and puts it away.

The toughness and endurance of the ass were celebrated in the popular story of the American steelman, Joe Magarac; the name "Magarac" is Slovak for "jackass." "All I do is eatit and workit same lak jackass donkey," said Joe. He was born from a mountain and had superhuman strength. When there was no more metal to mine, Joe stepped into the furnace and melted himself down into steel. This tale is often taken for folklore, but Owen Francis created the character in an article from the November 1931 issue *of Scribner's Magazine.*

The donkey is now most familiar as a symbol of the Democratic Party in America. The idea was partly inspired by the statement of Ignatius Donnelly to the legislature of Minnesota not long after the Civil War that the

Democratic Party was like a mule, lacking both pedigree and posterity. Thomas Nast popularized the symbol in political cartoons. The metaphor was first intended as an insult, but the Democrats certainly didn't mind. In fact, they adopted the symbol officially in 1874. Perhaps they realized that the symbolism of the donkey has always had many layers. If the long ears of a donkey suggest foolishness, its large teeth are a formidable weapon. The donkey is tough; he has a devastating kick. He may have a reputation for stubbornness, but isn't that often a virtue in politics?

Among the most eloquent tributes to animals ever written is *Platero and I* by the Juan Ramón Jiménez (1957). It is a series of remarks addressed by the author to a gentle donkey named "Platero," "loving and tender as a child but strong and sturdy as a rock." The donkey is not only a helper but a wonderful listener as well. The companions enjoy together the flight of butterflies, the playing of children, the touch of water, and all the richly sensuous life of a remote village in Spain.

As the horse took over the more glamorous role as the mount of warriors, the camel, like the ass, was increasingly regulated to bearing burdens. Though sometimes praised for humility, the camel had the additional reputation of being lascivious. Jeremiah used the camel as a symbol of Israelites who had commerce with heathens:

> A frantic she-camel running in all directions
> bolts for the desert,
> snuffing the breeze in desire;
> who can control her when she is in heat? (Jeremiah 2:23-24).

Though the Bible does not specify their mounts, the Magi or wise men that brought gifts to the infant Jesus are traditionally portrayed riding on camels, perhaps to indicate their exotic origins. The Wife of Bath in Chaucer's *Canterbury Tale*s, however, urged women to fight their husbands like camels.

The camel, in addition to sharing a reputation for foolishness with the ass, had the additional stigma in European culture of being filthy and ugly, a sort of deformed horse. Medieval bestiaries said that camels are so drawn to slime that they avoid clean waters for dirty ones. This ill repute even obscured the grace of the giraffe that, at least since Pliny the Elder through the Middle Ages, had the misfortune of being considered a camel with spots—a "cameleopard."

Advertisers in the mid-twentieth century exploited the reputation of the camel for ugliness by creating the cartoon character of Joe Camel, who had the face of a camel and the body of a man, to suggest a blue-collar toughness. He was used to sell Camel Cigarettes, which have the animal standing before a pyramid as their symbol, and both humps suggest the belly of a pregnant woman—a proof of his virility. Joe was such a commercial success that anti-smoking activists singled him out for protests, and ads featuring Joe made il-legal around the end of the nineties. Today, the camel of Arab lands, like the ass and mule of the Occident, has become a symbol of a vanishing way of life.

In Latin America, the llama was domesticated as a beast of burden about 5,000 years ago, approximately the same time as the donkey in the West. Statues of llamas have been found in tombs of the Mocha and Inca Indians, based in what is now Peru. The pre-Columbian figurines are often quite animated and individual, suggesting that llamas were viewed with a good deal of affection, though they may not have been accorded numinous powers. They still carry bundles for Indians high in the Andes Mountains, an environment to which they are very well adapted. At lower altitudes, the mule usually is the preferred helper of many solitary workers such as ped-dlers. Even today, in many villages, llamas and mules are used to deliver mail, and there is a quiet intimacy between these animals and their handlers.

HORSE

> Are you the one who makes the horse so brave
>> And covers his neck with flowing hair?
> Do you make him leap like a grasshopper?
>> His proud neighing spreads terror far and wide.
> Exultantly he paws the soil of the valley,
>> And prances eagerly to meet the clash of arms.

> —JOB 39:19-21

The experience of riding a horse may be the closest thing to union with an animal that people have ever known. The rider may determine the direction; the horse sets the rhythm. The rider rises and falls in a sort of trance as landscape passes beneath. The union of animal and man is commemorated in myths of the centaur, human to the waist and equine beneath.

For the most part, centaurs were dominated by animal instincts. Perhaps the fierce horsemen of the Eurasian steppes, glimpsed from a distance by people who did not yet know how to ride, inspired the legends of centaurs. The drinking and lechery of centaurs were notorious. Invited to the marriage feast of Pirithous, king of the Lapths, and princess Hippodamia, they became drunk and tried to carry off the bride. The Lapths finally defeated them in a ferocious battle, which has usually been interpreted as the triumph of refinement over barbarism. The story was a favorite subject of painters during the Renaissance.

There are several colorful myths about the origin of centaurs. In one tale, Ixion, King of Lapithea, was invited to Mount Olympus to dine with Zeus. He tried to seduce Hera, the wife of his host, and thought he had succeeded, but had actually only made love to a cloud. From that union, the centaurs were born. The tale is, perhaps, about the power of imagination, which centaurs have symbolized ever since. The story, however, does not have a happy end, for Zeus fastened Ixion to a fiery wheel that endlessly revolves in Hades.

Apollonius of Rhodes wrote that the deity Cronos was once having an affair with Philyra, daughter of Ocean, when his consort Rhea surprised him. Chonos leaped out of bed, changed himself into a stallion, and galloped off. Philyra wandered away in shame but later gave to birth to the centaur Chiron. Among centaurs, only Chiron had a reputation for great wisdom. His students included the warrior Achilles and the physician Asclepius.

Sometimes the wildness of centaurs proved more powerful than civilization. Ovid told the story of Canus, a young girl who made love to the god of the sea. When the deity promised to grant any wish she might have, she asked to become a man, so that she might take part in battle. The god granted her request, but, fearing for her safety, he added the gift of invulnerability to any weapon made of metal. Canus proved invincible in battle until she (or "he"?)

*The wise centaur
Chiron teaching the
young Achilles.*

*(Illustration by
Willy Pogany, 1918)*

encountered a band of centaurs. Not having learned to melt metal, they attacked and killed Canus with their primitive weapons such as rocks and sticks.

In the mythology of northern Europe, the giant horse Svadilfari was a sort of centaur. The Norse gods once commissioned the giant Fafner to build the walls of their city Asgard. If he could finish the work by spring, he would receive the goddess Freya in marriage, together with the sun and the moon; if not, he would forfeit all pay. Svadilfari helped the giant by not only drawing the stones but also setting them into place. Loki, the mischievous god of fire, changed himself into a mare. Svadilfari ran after Loki, so the wall remained unfinished. From the union of Sadilfari and Loki, the eight-legged steed Sleipnir was born, on which the god Odin journeyed through the sky and to the realm of the dead.

In the time of Homer, horses drew chariots but did not yet carry riders on their backs. After the Trojan War had dragged on for ten years, the Greeks pretended to sail away and left a giant horse made of wood on the shore. The Trojans, thinking the horse an offering to the gods, took it inside the city walls. The Greeks had only temporarily withdrawn to a nearby island, and had concealed warriors inside the wooden image. When night came, they slipped out and opened the city gates for invaders. Apollodorus wrote that Helen of Troy had walked around the horse imitating the voices of the wives of men inside. One of the men wished to call out in reply, but Odysseus clamped his hand over the soldier's mouth. Perhaps this incident records an imperfectly remembered rite of fertility, an offering to a horse deity that the Greeks believed had granted them victory.

Horses have been found in graves in France, Ukraine, Scandinavia, China, and many other places, buried to carry their masters in the world to come. Shamans, especially in the Arctic Circle have sacrificed horses to accompany them on ecstatic journeys to other worlds. Apollo, the Roman god of the sun, rode across the sky in a chariot drawn by horses; so did the Persian Mithras. Perhaps the most popular symbol of transcendence today is Pegasus, a horse with wings, which sprang from a drop of blood of the monster Medusa after she was decapitated.

Though primarily herbivorous, horses have always been closely associated with war. Plutarch, in "Of Isis and Osiris," told an Egyptian legend that Osiris, god of the dead, was once instructing his son Horus. The father asked what animal would be most useful in battle, and Horus replied that it was the horse. Osiris asked why his son had not picked the lion, to which Horus replied that, "a lion was a useful thing for a man in need of assistance, but that a horse served best for cutting off the flight of an enemy and annihilating him." On hearing this, Osiris realized that his son was ready to become a warrior.

Poseidon, the Greek god of the sea, was also a deity of horses, and sired the first horse. According to one myth, he moistened a stone with his semen (that is, the foam of a wave), thus fertilizing the earth, and out leaped the horse Scyphius. In another, Poseidon, in equine form, raped Demeter, the goddess of grain, after which she gave birth to the horse Arion. Hippocamps—with the heads of horses, the bodies of serpents, and the tails of fish—drew Poseidon's chariot. Horses rise and fall like waves as they gallop along, while their manes are like foam. In a herd, they move to the rhythm of flowing water.

When Poseidon and Athena, the goddess of wisdom, were competing to be patron of the largest city in Attica, the immortals decided that the role would go to whoever gave the greatest gift. The god of the sea struck the ground with his trident and made a horse spring forth. Athena, however, created an olive tree, and her gift was judged finer, so the city was named "Athens" after her. The contest records a conflict between the nomadic life, in which the horse is absolutely central, and the settled life of agriculturalists.

In Greek mythology, King Diomedes of Thrace fed strangers who visited his land to his mares. The ninth labor of Hercules was to capture the man-eating monsters. After a terrible struggle, he killed Diomedes and fed the King's lifeless body to these horses, which, after eating their former master, returned to a normal equine diet. The tale reflects the fear that horses inspired at a time when they were not yet fully domesticated but might still be seen roaming in the wild. The Romans knew horses only as domestic animals, but they admired the way they remained spirited even in the service of human beings.

In the latter fourth century BCE, the Greek mercenary Xenophon wrote the first book on horsemanship that has come down to us. He approached horses without awe but with remarkable concern and respect. The horse, he understood, was the soldier's companion in battle and shared his hardships. Much of his treatise was devoted to detailed descriptions of such matters as how to place the halter about the neck of a horse. Xenophon's foremost rule was never to deal with a horse thoughtlessly or in a fit of anger.

The Roman Emperor Caligula talked of naming his favorite horse Incitatus as consul. The most popular sport in ancient Rome was chariot racing. The Romans would race two chariots, each drawn by two horses, at the festival of Mars in the middle of October. The finest of the horses, the one who drew the winning chariot on the inside of the track, would then be sacrificed to Mars, the god of war. The head and tail of the horse would then be cut off and decorated. Sometimes, the head would be fixed to the house of a prominent farmer or other citizen.

The horse was also the only animal to set apart an entire social stratum—that of "equestrians" or horsemen in Rome, which eventually became the knights of the European Middle Ages. This was an elite class of warriors, for riding was confined to officers in the Roman army. Finally, however, their superior horsemanship enabled barbarian tribes to conquer Rome. In contrast to the Romans, who made riding a privilege reserved for the upper class, the tribes had entire mounted armies. The Roman historian Ammianus Marcellinus, writing in the fourth century CE, described the Huns as primitive, bestial people who were almost one with their horses; they would not dismount even to eat or sleep.

Until the two world wars, battles were very often won or lost by gallant cavalry charges. The Biblical prophet Elijah ascended to heaven in a fiery chariot drawn by horses (2 Kings 2:11). The horse is also one of twelve animals in the Chinese zodiac. The last incarnation of Vishnu, when he comes to bring salvation to the world, is to be the horse Kalki. Mohammed rode a horse named Al-Borak from Mecca to Medina and up to heaven. Al-Borak, who was generally painted with the face of a woman, could predict danger and see the dead.

The terrifying Russian witch Baba Yaga may originally have been a horse-goddess. She lived in the depths of the woods in a house on chicken legs, surrounded by a fence made of human bones that were topped with skulls. She ate people like chickens, but the heroes in fairy tales often risked death by visiting her in search of help through magic or sage advice. Baba Yaga had a herd of mares, which were her daughters, and sometimes Baba Yaga herself became a mare.

In the story of "María Morevna," recorded by Alexander Afanas'ev, Prince Ivan went to Baba Yaga and asked for a horse fast enough to escape Koschei, the spirit of death, who had stolen his bride. He obtained his goal by tending the horses of Baba Yaga, with the help of friendly animals, for three days. Riding the horse given to him by Baba Yaga, Prince Ivan rescued his bride. His beloved, María Morevna, was a spirit of vegetation who had

to spend part of the year in the underworld, much like the Greek deity Persephone. Koschei was a ruler of the Underworld, similar to the Greek Hades
or the Roman Pluto. Baba Yaga, though far less benevolent, has some of the
traits of Demeter, mother of Persephone, who was also mistress of horses.

A similar myth may be the origin of the famous English story of Lady
Godiva. Lord Leofric had taxed his people beyond what they could bear.
His wife Godiva spoke in their defense, and he agreed to reduce his taxes
only if she would ride naked through the streets. She did so, covered only
with her long hair, and all the people turned their heads away, so as not to
look, except one man named Tom, who then immediately went blind.
Leofric, in shame, revoked all taxes except those on horses.

The horse was also sacred to Freya, the Norse goddess of love and fertility, who may be the origin of Godiva's legend. An image of that goddess
had once been drawn through the streets by horses every year in spring. Like
Persephone, Maria Morevna, and Freya, Godiva may have been a personifi-
cation of vegetable life, while Leofric was a spirit of winter. "Peeping Tom"
may originally have been a human sacrifice accompanying the rites of Freya.

Horses had been often associated with fertility, so, as the story of Lady
Godiva illustrates, the intimacy of horse and rider seems to have a sexual
side. People worried that grooms might mate with horses, producing who-
knows-what dreadful progeny, and such bestial unions were often punish-
able with death in the late Middle Ages and Renaissance.

Since horses were especially associated with masculine power, women
had to be especially careful. Modesty required that they not place their legs
around the animal. On Roman coins, the Celtic horse-goddess, Epona, was
displayed riding sidesaddle upon a mare. Thomas More, who had a ribald
as well as a saintly side, once rebuked his daughter who did not wish to ride
sidesaddle as befitting a lady, by saying, "Well, my girl, no one can deny that
you are ready for a husband, since your legs can straddle so large a horse."
In Shakespeare, when the young Venetian girl Desdemona has eloped with
the Moor Othello, the villainous Iago says to her father, " . . . you'll have
your daughter covered with a Barbary horse; you'll have your nephews
neigh to you; you'll have coursers for cousins . . . " Enemies of the Russian

Empress Catherine spread rumors, which circulate to this day, that she copulated with her horses. The horse was often considered a disguise for the devil or a companion for a witch.

The sexuality of horses was sacralized in the figure of the unicorn. This fantastic animal had, in the course of a long intricate history, also taken on features of the ass, rhinoceros, narwhal, goat, and many other creatures. In Medieval Europe, the unicorn was thought of primarily as a horse with a single horn. That protuberance, clearly phallic, seemed to absorb and transform all that was disturbing in equine sexuality, leaving the animal divinely chaste. According to legend, the unicorn could not be caught by force, but it would come to a young virgin and lay its horn in her lap, at which point it might be captured without difficulty. This story reflected archaic mythology of the sacred hunt, as the young girl came to symbolize Mary and the unicorn to symbolize Christ. The capture and slaying of the unicorn were an allegorical crucifixion acted out not with malice but with solemnity in the tapestries and paintings of the Middle Ages. Nobody has entirely reconstructed the meaning of this elaborate allegory, but it may be about domestication. Our sexuality, like the unicorn itself, cannot be overcome through force alone, yet, also like the unicorn, it can be tamed.

Archaic beliefs about horses show up frequently in European fairy tales. In the fairy tale "Faithful John" by the Grimm brothers, ravens tell a servant that a horse will trot up to the young King, to take him up into the sky forever. The King may only be saved if somebody jumps on the horse and shoots it quickly. This probably alludes to an archaic horse sacrifice that accompanied many shamanic journeys. In Grimms' "The Goose Girl," a princess speaks to the head of her beloved horse that has been sacrificed. The animal offers sympathy but no sage advice, showing that the old magic will not work in the modern world.

Many Catholic saints are patrons of horses, no doubt a legacy of pagan times. One is Saint Eloy, who was a blacksmith to the Frankish kings. Once, he was asked to shoe a horse possessed by the Devil. He blessed the animal, took off its legs, placed shoes on its feet, and then returned the limbs to their rightful place. Another patron of horses is Saint Stephen, the Church's first martyr. On his feast day, December 26, parishioners in Poland shower the priest with oats as a gift from their horses. But horses, like most animals that had once been sacred, were often demonized in the late Middle Ages and Renaissance. The long association of horses with the poles of fertility and death did not always make them beloved, nor did their longstanding association with goddesses.

Horses became familiars of witches and, at times, even a disguise for the Devil himself. One example comes from *The Quest of the Holy Grail*, an anonymous Medieval romance from Britain written in the early thirteenth century. Sir Perceval, the holy fool of Arthurian legend, found himself alone in a vast forest. A woman appeared mysteriously and offered him a horse in exchange for his service, and Perceval accepted with delight. She went for a moment into the depths of the woods and returned leading an enormous black horse. Perceval could not look on the horse without fear, but he mounted it and galloped on until the trees ended and a wide river crossed the path. No bridge was to be seen yet the horse continued onward. Unable to stop, Perceval made the sign of the cross. At that instant, the horse threw him to the ground and plunged into the river. Perceval heard howling and shrieking, as flames rose from the water. The woman and the horse had both been devils.

The equestrian, or knightly, tradition lasted longest in the American continents, where there was plenty of open space, and large herds of feral horses roamed the plains of the United States and Argentina. In stories of the American "Old West," wandering cowboys were a bit like the knight-errants of the European Middle Ages, moving from one town to the next in search of fortune and adventure, but the role of horses in battle was steadily reduced as the world industrialized. The disastrous, if valiant, charge of the Polish cavalry against German tanks at the beginning of World War II has become a symbol of romantic resistance to the industrial world.

Horses gradually came to be associated more with recreation than with work, and more with women than with men. The novel *Black Beauty* by Anna Sewell (1877) started a tradition of horse stories primarily for young girls. The book is a fictive autobiography of a stallion that draws carriages, starting with his early life in open meadows, going through several masters, and ending with a pleasant retirement on a farm. Horses now represent a gentle aspect of male sexuality, which contrasts with the machismo that is so often celebrated in the mass media of today.

Today, horses are also becoming nostalgic reminders of our past. The horse race, that favorite sport of the ancient Romans, is now a gathering point for many who seem in some way bypassed by the modern world, from aristocrats to Mafiosi. Horses, however, are still a common means of transportation in remote areas where there are few roads. Their maneuverability still makes them helpful to police, even in cities, and horses have been trained to enter apartments and walk up stairs. But urban horses are rare enough to make many bystanders stare in surprise and admiration whenever they appear in public.

Though horses are far slower and less powerful than our motors today, the horse still provides the ultimate model for our locomotives, automobiles, and rocket ships—for anything that is fast and sleek. The word "horsepower" is still used as a unit of force in motors. The metaphor of the "Trojan Horse" is perhaps more widely used than any other image from the ancient world. It can refer to just about anything from internal subversion in a government to a computer virus. In the late 70's a nuclear power company was called "Trojan" and had a horse as its logo. Demonstrators against nuclear power in California built a large wooden horse. As they paraded it around, out popped a person dressed as the Grim Reaper with his scythe.

11
NOBLE ADVERSARIES

THE CAVE PAINTINGS OF PALEOLITHIC EUROPE, TOGETHER WITH artifacts and possible parallels with other cultures based on hunting, suggest a solidarity with animals that continues to intrigue people today. The mentality that produced them is, however, impossible to reconstruct them with any confidence. We do not know, for example, whether the people thought of the shedding of blood as a transgression, which needed to be assuaged through ritual. We do not know whether hunting was accompanied by feelings similar to what we now call "guilt." However that may be, some scholars of myth have attempted to trace not only the origins of religion but even storytelling itself to the hunt. Many contemporary sportsmen continue to use elaborate rituals, but are often frustrated at the difficulty of communicating the meaning of the hunt to outsiders. Their identification with the hunted animal, though it may seem paradoxical, can be very intense. In some cases, such as that of the Great Plains Indians, this totemic bond can be the foundation of a culture.

HART AND HIND

> O world! Thou wast the forest to this hart;
> And this, indeed, O world! The heart of thee.
> How like a deer, stricken by many princes,
> Dost thou here lie!
>
> —WILLIAM SHAKESPEARE, *Anthony and Cleopatra*

Just as we think of the male lion in terms of his luxurious mane, we think of the stag in terms of his enormous antlers. These also define the female, the deer (or hind), through their absence or, in some species, their relatively small size. Together hart and hind represent nothing less than the primordial division into male and female. Through this symbolism, they are also intimately associated with the forests where they live. They move among the trees silently, blend in perfectly, and have a way of suddenly appearing, which makes them seem to be the very soul of the woods.

Most of the time, we use the word "deer" as a plural to include both male and female, suggesting that the species is at least primarily feminine. The English word "deer" originally meant "animal" in general, as opposed to humankind, just as its relatives such as "tier" (German) and "dier" (Dutch) still do. The current meaning of the word in English was only established in the 15th century. This etymology suggests that people thought of deer as a sort of epitome of what it means to be a beast of the forest and field—our quintessential Other.

Why was the deer not domesticated in Neolithic times, along with the sheep, goat, pig, and chicken? It had already long been a major source of food for human beings, and is fairly easily tamed. The answer of anthropologist Betrand Hell is that leaving the deer free to roam the forests was a deliberate, if not necessarily conscious, part of the social construction of the domestic and wild realms. The deer would embody the forest as a realm of wonder, danger, magic, and adventure. Human beings needed the hunt, and its mystical solidarity with the animal world, more than they craved an easy source of protein.

One of the earliest possible mythological creatures is a muscular, bearded human figure known as "the sorcerer," bearing antlers of a stag, painted over a millennium before Christ in the French cave of *Trois Fréres*. His head is disproportionately large, and the lines of his body are sharply angular. Perhaps he is doing a sort of dance. This figure could be either a nature spirit or else a shaman who wears the antlers and mask of a stag. For those who created the image, the two alternatives were perhaps not so very far apart, since the spirit of a stag could have possessed a shamanic dancer.

No creature is more quintessentially masculine than the stag, with his large shoulders, impressive size, athletic stride, and frequent propensity to fight. The doe seems comparably feminine with relatively modest size and cautious, delicate steps. Furthermore, does tend to remain in a herd, suggesting the traditionally feminine role of women as guardians of the home. Stags, by contrast, tend to be solitary except during mating season. One may choose to call the contrast either "archetypal" or "stereotyped," but it is hard not to think of the division of roles along lines of gender in many human societies.

This solitary life is why the stag has been associated with many holy anchorites. Saint Hubert, for example, was once a pleasant but shallow courtier at the court of King Pepin in France. His great passion was for the chase, and he even skipped church service on Good Friday to hunt. When he finally caught sight of the stag, he saw a vision of the crucifix between its antlers, and the voice of Christ warned Hubert that he must either accept God or end up in Hell. Hubert repented his frivolous ways, became a hermit, and devoted himself to God. He is now the patron of hunters, and a festival in his honor is still celebrated by them every year in the forests of France.

Christ also took the form of a stag in visions of Saint Pacidus and Saint Eustace, about whom very similar tales are told. The anchorite Saint Giles once sheltered a deer from a hunting party of the Emperor Charlemagne, and an arrow intended for the stag hit the saint instead. In gratitude, the deer brought him milk every day, while the King built a monastery in his honor. Saint Ciaran, and early disciple of Saint Patrick, had several deer among his earliest disciples when he became a missionary in Ireland. A stag accompanied Saint Canice, an Irish hermit, and offered the holy man his antlers as a bookstand.

The stag has not only been a common symbol of Christ, the resurrected God, but it also represented the entire natural world, subject to cycles of germination, growth, death, decay and resurrection. Deer, like plants but unlike most animals, clearly reflect the seasons in their appearance. Their fur acquires a richer color in summer. Even more significantly, the males of many species shed their horns every year. Up through Elizabethan times, it

Shou Hsing, a Chinese god of longevity, mounted on a stag.

was widely believed that the sex organs of stags as well were shed and rejuvenated every summer. The lifespans of large trees were often attributed to stags. One popular Medieval legend concerned a stag that was given a golden collar by Alexander the Great (Caesar, Charlemagne, or King Alfred in other versions). This animal was reportedly found hundreds of years later, still fully vigorous.

Tradition grants even greater longevity to deer in Asia. According to Chinese tradition, deer can live over 2,500 years if never killed. During the first 2,000 years, the skin of the deer gradually turns white; its horns turn black in the next 500. From that time on, the deer ceases to grow or change, though it may subsist on only clear streams and lichens in the mountains.

The identification of the stag with the forest was so intimate that it guided scientific opinion until around the end of the eighteenth century. The horns of the stag were, throughout the Middle Ages, understood as the branches of a tree, the skin covering them as a kind of bark. Georges Luis LeClerc de Buffon, a foremost zoologist of the eighteenth century, theorized that, since the stag nibbles so much on trees, branches eventually grew out of its head.

The association with the forest is partly responsible for another important feature of the stag in folklore. While the earth is traditionally perceived as feminine, we generally view trees, with a few exceptions, as masculine. The ash was sacred to Odin, the oak to Zeus and Thor. Deer have frequently been depicted around the base of a tree of life in ancient Mesopotamia and several other regions. According to Norse mythology, stags nibbled at the base of the world tree Yggdrasil.

The horns of the stag have been worn by many highly masculine heroes and gods in folklore and mythology, such as the Celtic god Cernunnos. Stags sometimes drew the chariot of Dionysius, the Roman Bacchus. Neo-pagans of today celebrate the stag as a symbol of male energy, and, by extension, the god. In European folklore, the sexuality associated with this animal is fierce and unrestrained, yet also completely chaste. Neither fundamentally romantic nor promiscuous, it is close to the undifferentiated sexuality of the vegetable realm than that of either other animals or human beings.

By about the time of Charlemagne, the hunt of the stag was a "sport," a privilege of princes and aristocrats that was forbidden to the peasantry, often on pain of death or mutilation. It was set apart from the everyday world by elaborate costumes, vocabularies, and rituals. A solemn rite, reminiscent of Holy Communion, followed the slaying of the stag. The body of the stag was divided among participants, from the dogs to the horsemen, and the head was finally presented to the lord of the manor.

A doe traditionally accompanies Artemis, goddess of the hunt and protector of animals. King Agamemnon, commander of the Greeks in their assault on Troy, once killed a hind sacred to Artemis, boasting that he was greater than she in the hunt. The goddess stopped the winds, thus forcing the Greek ships to remain in port. An oracle told the Greeks that there would only be wind when Agamemnon sacrificed his daughter Iphigenia to Artemis. Under pressure from the other Greek princes, Agamemnon finally complied. When Iphigenia was led to the altar, Artemis substituted a hind and carried away the young girl in a cloud to Tauris on the island of Crimea, where she became a priestess.

In his *Metamorphoses*, Ovid tells how the young hunter Acteon once came upon the goddess Diana, bathing with her nymphs at a stream. He was so overwhelmed by the sight that he stood transfixed, until the goddess became aware of him and shouted, "Tell people you have seen me, Diana, naked! Tell them if you can!," as she splashed water in his face. Acteon looked into the water and saw, with horror, that he was turning into a stag, and then he was torn apart by his own hounds. Some interpreters believe that Diana here represents Julia, daughter of the Emperor Augustus, while Acteon represents Ovid himself, who was banished for witnessing one of her affairs. The unfortunate hunter may also be a man who intrudes on women's mysteries. For alchemists of the Renaissance, Acteon was the adept who, on understanding the nature of reality, must die in order to be reborn.

The deer was also sacred to many other archaic goddesses including the Mesopotamian Ninhursag, the Egyptian Isis, and the Greek Aphrodite. The nurturing quality of a hind is shown in the Greek myth of Telephus, son of Hercules and Auge. His mother, to conceal her affair, hid the infant Telephus in the temple of the goddess Athena, which caused the local harvests to fail. When this was discovered, the infant was exposed on a mountain to die, but he was saved by a hind that suckled him.

The Old Testament often celebrates the hind as a symbol of feminine virtue, for example:

> Find joy with the wife you married in your youth,
> Fair as a hind, graceful as a fawn.
> Let hers be the company you keep,
> Hers the love that ever holds you captive (Proverbs 5:19).

Psalm 42 begins "As the doe longs for running streams, so longs my soul for you, my God."

Stag and hind share an importance as divine messengers throughout most of Europe. In countless epics and fairy tales, deer appear to the hero and guide him or her through the woods. The search for the Holy Grail begins as knights of King Arthur follow a white stag into the woods. According to legend, Louis the Pious, son of Charlemagne, was once so intent on pur-

The stag that took pride in his horns found they got caught in the trees and prevented his escape from the hunters, in the Aesopian fable.

(Illustrated by J J. Grandville, from Fables de la Fontaine, 1839)

suit of a stag that he became separated from his hunting party, fell from his horse while crossing a stream, and became lost in a vast forest. As his men found him next morning, daylight revealed a bush or roses blossoming in the snow, a miraculous sign that he should build a cathedral in that place.

In Celtic myth, deer are guides to the Other World, while in the lore of Ireland they are also sometimes the cattle of fairies. The deer hunt also provided a frequent metaphor in Medieval and Renaissance culture for courtship. Imagery taken from the chase was used to express the fear, longing, concealment, and tension that are often part of erotic intrigues. In Shakespeare's *Twelfth Night*, Curio tried to divert the melancholy Duke Orsinio:

> CURIO: Will you go hunt, my lord?
> DUKE: What, Curio?
> CURIO: The hart.
> DUKE: Why, so I do, the noblest that I have.
>   O, when mine eyes did see Olivia first,
>   Methought she purg'd the air of pestilence!
>   That instant was I turn'd into a hart,
>   And my desires, like fell and cruel hounds,
>   E're since pursue me.

Even more frequently, the woman was viewed as the hunted deer. The horns of a stag were often used as a symbol of cuckoldry, an obsession in literature of the Elizabethan Age.

But for the more humble social orders, the deer were often bitterly resented as a symbol of oppression. In the stories of Robin Hood, the merry men living in Sherwood Forest constantly defied the Sheriff of Nottingham by feasting on the "king's deer." When the prohibitions against hunting deer by the peasantry were finally lifted, many commoners asserted their new freedom by pursuing the chase with enormous vehemence and cruelty. The British pastor Gilbert White complained "Unless he was a hunter, as they affected to call themselves, no young person was allowed to be possessed of manhood or gallantry."

In *Bambi, A Life in the Woods* (1928), Felix Salten protested this indiscriminate hunting in his beloved Vienna Woods. The novel recounts the coming of a young stag named Bambi from his birth in a clearing. He learns the ways of the woods, fights off his rivals and mates with a hind named Filene. The tale ends with the death of Bambi's father, at which point Bambi goes off by himself to become a guardian of the forest. The poacher is a figure of awe and terror who killed Bambi's mother and allowed the deer no rest. Salten, himself an avid sportsman, later tried to draw a sharper distinction between the poacher and the legitimate hunter in a sequel. The enormously popular movie by Disney Studios that was based on the book (1942), however, further anthropomorphized Bambi and Filene, while it made "man," known only through fires and bullets, seem like a brutal, sinister force of nature.

The close association of deer with the landscape is also found in the cultures of Native Americans. Several tribes of Mexico and the American Southwest including the Aztecs, Zuni and Hopi traditionally perform a deer dance to influence the elements and bring bountiful crops. The Yaqui Indians today generally follow a Catholicism that has blended with their traditional, tribal religion, and they have incorporated the deer dance into an Easter ritual that is performed every year. In all of these ceremonies, the chief dancer will wear either the head of a deer or antlers and will imitate the motions of the animal, which looks cautiously about while moving through the woods.

With the increasing suburbanization of North America and Europe, many deer have lost their fear of human beings. White tailed deer of North America, which live on browse at the forest's edge, are now actually far more common than they were in the time of Columbus. Many suburbanites consider deer "pests" or "overgrown rats," complaining that they cause traffic accidents or nibble at gardens. Perhaps even more seriously, they can prevent the regeneration of forests by eating young shoots. Perhaps it is only in the far North, where many of the few remaining wildernesses may still be found, that the ancient symbolism of the deer, particularly the moose, as guardians of the wood retains much of its vividness.

AMERICAN BUFFALO

"What is life? It is the flash of a firefly in the night. It is the breath of a buffalo in the wintertime."

—CROWFOOT

In *Animals of the Soul*, Joseph Eppes Brown writes that for the Oglala Sioux, "the buffalo (or "bison") is the chief of all animals and represents the earth principle, which gives rise to all living forms." Black Elk, the shaman who initiated him into the ways the Sioux, said that that hunting was a "quest for ultimate truth." In the culture of the Indians of the Great Plains, the buffalo has a significance that seems, in many ways, to parallel that of the deer among Europeans. Both animals were hunted amid elaborate ritual and ceremony, and were often associated with visions. The buffalo embodies the prairie for the Indians, just as the deer embodies the forest for Europeans. Despite giving an impression of wildness, the forests of Medieval Europe were mostly preserves that were carefully managed around the needs of the deer, which would be hunted by royalty and nobility. In a similar way, Indians had managed the prairies around the buffalo by deliberately starting fires, destroying young trees and rejuvenating the grasses.

But these similarities resulted less in sympathy than in misunder-

When this was painted, such hunts had long since disappeared, but were remembered with nostalgia by both settlers and Indians.

(Buffalo Hunt *by William R. Leigh, 1947*)

standing. For Europeans of the early modern period, hunting deer had long been an exclusive privilege of inherited wealth and power, and others who engaged in it were subject to terrible punishments. They could little comprehend that for the Indians, the hunt was a livelihood as well as a religious act. For the democratically minded emigrants to ~~the~~ New York, the Indian hunters seemed to be indulging in a life of idleness, much like the monarchs and nobles they had come to despise.

In pre-Columbian times, hunting buffalo on the Great Plains was a perilous enterprise. Occasionally, Indians were able to frighten a herd into stampeding off a cliff. More often, they would dress themselves in buffalo or wolf hides, at times mimicking the sounds and movements of those animals, as they waited for an auspicious time to strike. With horses imported by Europeans, hunting became much safer and more efficient, but that did not diminish the religious significance of the hunt. Perhaps the shared experience of galloping across the plains helped the Indians to identify more closely with the hunted buffalo. In the early twentieth century, Black Elk also told that his people had once been visited by a buffalo goddess in the form of a beautiful lady, the White Buffalo Calf Woman, who gave them a sacred pipe carved with the image of a buffalo calf, which is the center of their rituals. The Sioux, as well as the Pawnee, believe that a cosmic buffalo stands at the gate through which animals pass as they die and are reborn.

When the great westward migration of European Americans greatly accelerated at the end of the Civil War, pioneers passed through, and often tried to claim, what had been Indian territories. The United States military was unable to decisively defeat the Indians militarily until they began to deliberately destroy the buffalo herds, thus simultaneously attacking the Indians' source of food and their religious beliefs. In 1850, there had been an estimated fifty million buffalo on the prairies; by 1900 only twenty-three were left in the United States and about 500 in Canada, and the Indians had been forced to surrender most of their lands. The prairies themselves had been transformed, often into pastures for European cows. Many Indians believe that the buffalo had sacrificed themselves and died in place of the Indians.

In the twentieth century, attempts have been made to preserve the buffalo in the United States. The largest herd is that in Yellowstone National Park, which contains a few thousand animals, and a few of them are now being transferred to tribal reservations. Many Indians still regard them as ancestors. For most Americans, Indian or not, the buffalo remain a symbol of the expansiveness and freedom of the open plains, in a mythic time before the innocence of the nation was lost. From 1920 through 1938, the treasury of the United States issued the Buffalo Nickel, with a buffalo on one side and the profile of an Indian, accompanied by the date and the word "liberty," on the other. In 1994, a white buffalo calf named "Miracle" was born in Janesville, Wisconsin, and many Sioux elders interpreted the birth as the return of the White Buffalo Calf Woman, heralding the start of a new era.

The Badger and the Mole. Illustration by Arthur Rackham for
The Wind in the Willows.

12
TOUGH GUYS

WE THINK OF THE URBAN ENVIRONMENT AS THE HUMAN REALM, while the natural world is for animals. In consequence, we credit animals that can thrive in towns and cities—creatures like rats and pigeons— with a special toughness, a bit like the people who grow up in the most dangerous areas of our inner cities. That is not entirely mistaken, but we should remember that many of these animals have been thriving in cities since they were first built—in other words, as long as people. Many of them have lived off food that people discard or waste. They are now as thoroughly adapted to urban environments as human beings are. At times, they are resented as "parasites," but that is generally unfair. Except possibly for rats, mice, and a few insects, they live in a symbiotic relationship with human beings, and perform a valuable service by disposing of organic material that would otherwise decay.

BADGER, ERMINE, GROUNDHOG, AND SQUIRREL

Come play with me;
Why should you run
Through the shaking tree
As though I'd a gun
To strike you dead?
When all I want to do
Is to scratch your head
And let you go.

—W. B. YEATS, "To a Squirrel at Kyle-na-no"

Though usually comparatively small, weasels do not hesitate at all to do battle with rats or snakes. Pliny the Elder stated that the weasel was the only animal that could defeat the basilisk, a serpent able to kill other creatures with a single glance. The diminutive creature overcoming such a monster later became a symbol of Christ triumphing over the Devil. In the modern era, as traditional martial virtues have come to be less valued, the weasel's reputation has declined. In Kenneth Grahame's *The Wind and the Willows* (1906), weasels form a vicious yet cowardly mob.

The ermine, a relatively large weasel, was sacred to the Zoroastrians. Because of its white color, it has often been associated with the fierce chastity of a soldier of God; Mary Magdalene was depicted wearing an ermine coat in token of her reformation. A popular European legend stated that an ermine, pursued by hunters, would allow itself to be killed rather than soil its beautiful coat with mud. Ermine fur is synonymous with luxury, and was worn by Louis XIV of France and other monarchs.

A related animal, even more important in folklore, is the badger, which is noted for its powerful front legs and long claws adapted for digging. Its burrow under the earth and nocturnal habits make the badger a creature of mystery. In China and, most especially, Japan, badgers are shape shifters, and many stories are told of spirits haunting old buildings, desolate fields or ponds, that turn out to be badgers. Typical is the story of an ascetic hermit on Mount Atago near Kyoto. A hunter would bring him food every day, and one afternoon the hermit confided to his benefactor that the Bodhisattva Fugen visited him every evening upon a white elephant. At the invitation of the hermit, the hunter stayed to see the Fugen. At first, the hunter was dazzled by the vision, but as he gazed more closely he began to feel suspicious. Finally, he shot an arrow at the vision. The Bodhisattva immediately disappeared, and there was a rustling in the bushes. "If it had really been Fugen," the hunter told the hermit, "the arrow couldn't have done any damage, so it must have been some monster." The next morning the two followed a trail of blood and found an enormous badger with an arrow through its breast.

The badger, actually a member of the weasel family, is often thought of as a small bear, and it is one of many animals that have taken the place of the bear in forecasting the coming of spring. The end of the winter was once indicated by the return of the bears from hibernation. As these large animals became scarce, their role as the heralds of spring was taken over in Germany and much of Britain by the badger. According to tradition, they emerge on the same day as the Roman Catholic and Eastern Orthodox festivity of Candlemas, which celebrates the presentation of Jesus, "light of the world," in the temple. According to a German proverb, "the badger peeps out of his hole on Candlemas Day, and, if he sees the sun shining, he draws back into his hole."

In the United States, the woodchuck or groundhog has replaced the badger in forecasting spring. On February 2—Candlemas Day, but now known popularly as "Groundhog Day" in the United States—the animal will reportedly lift its head out of its burrow. If it sees a shadow, the groundhog will return, and winter will linger six more weeks, but, if it stays out, spring is at hand. Newscasters gather around the homes of famous groundhogs, most notably Punxsutawney Phil in Pennsylvania, and film the moment of ascent. The result is predetermined, or at least manipulated, by heaters that are placed under the ground.

Nobody really pretends to take the augury seriously, but it does have very exalted roots in history and myth, going back to rural societies, in which agriculturalists watched for subtle signs such as the migration of birds or the emergence of animals from hibernation to decide on the best times for planting and harvesting. The slightly embarrassed laughter that usually accompanies the celebration of Groundhog Day may go back to Protestant mockery of Catholic rituals. With irony, and even a little sneering, Groundhog Day celebrates a rural way of life, which we can now barely remember, yet somehow miss. Holidays are times set aside to reflect on the meaning of our past, and Groundhog Day, while seemingly trivial, turns out to have a history as old and complex as perhaps as any celebration.

Especially beloved in the northern hemisphere, however, is the squirrel. It is primarily the long bushy tail that differentiates squirrels from rats, yet what a difference that makes in the way the two are regarded! Rats are often feared and despised, yet squirrels are such a part of our yards and parks that these places would appear desolate without them. Nevertheless, squirrels have been the subject of many ambivalent legends. For the Ainu of Japan, they represented the discarded sandals of the god Aioina, which would never rot, perhaps because squirrels move in spurts that are like foot-steps—they seem to hop more than to walk. Malaysians believed squirrels were produced, like butterflies, from the cocoon of a caterpillar, and they considered the dried penis of a squirrel a powerful aphrodisiac. In Norse mythology, the squirrel Ratatosk was the bringer of rain and snow. It moved up and down the tree of life, Yggdrasil, constantly trying to stir up strife be-tween the eagle at the top and the serpent at the base. In Irish mythology, the goddess Maeve has a bird perched on one shoulder and a squirrel on the other, her messengers for the earth and sky. The habit of hoarding nuts made squirrels symbols of avarice in some Medieval bestiaries, but Victorian books of natural history often praised squirrels for their thrift.

Today, squirrels entertain urban dwellers with their spectacular leaps between trees or by running along telephone lines over busy streets. The hero shows this fearlessness in Beatrix Potter's The *Tale of Squirrel Nutkin* by constantly, and unnecessarily, taunting an owl. Eventually, Nutkin is caught, and the owl nearly skins him alive, but he escapes after losing only a part of his tail.

In Britain today, the indigenous red squirrels are now endangered, due in part to diseases carried by grey squirrels, which were imported from North America. For many, the eradication of the grey squirrel has become a nationalistic cause, a matter of preserving British integrity from the cul-tural imperialism of the larger countries across the ocean. Since most of the remaining red squirrels are in Scotland, they are sometimes used as a symbol of Scottish nationalism. The red squirrel is a unique and very attractive an-imal, distinguished by large tufts of hair on its ears, and a very rich, reddish brown, color. All the same, it seems unfair to both red and grey squirrels, when they are made the focus of human disputes, in which they themselves have no part.

Every now and then, a squirrel will turn and stare at a person, with a gaze that suggests curiosity but neither fear nor anger. Because they seem completely untroubled by human presence, they reassure us that perhaps we have not alienated ourselves too much from the natural world after all.

FLEA, FLY, AND LOUSE

> *Little Fly*
> *Thy summer's play*
> *My thoughtless hand*
> *Has brush'd away*
> *Am I not*
> *A fly like thee?*
> *Or art not thou*
> *A man like me?*

—WILLIAM BLAKE, "The Fly"

The authors of the ancient world generally did not distinguish sharply between the different types of small insects which might be a minor if persistent irritation, and the term "fly" is used here loosely as a general designation for them. In the Biblical book of Exodus, the fourth plague sent by Yahweh when the Pharaoh refused to release the Israelites was a plague of gadflies that filled the palaces, a particularly insulting punishment since these insects are generally are attracted to cattle. The Egyptians themselves, however, seem to have admired the appearance of houseflies, which they frequently used in decorative pins. Pendants of gold in the form of flies were awarded to soldiers for valor.

In the play *Prometheus Bound* by the tragedian Aeschylus, Hera changed the maiden Io into a heifer as punishment for having an affair with Zeus. Then the goddess sent a gadfly to drive the unfortunate creature across Europe and Asia. A similar image is used, though in a positive way, in Plato's "Apology," where Socrates compared himself to a gadfly sent by God to prod the Athenians out of their complaisance. In a similar spirit, the Greek poet Melegros called on a mosquito to go and buzz in the ear of his beloved to remind her of his love. In many cultures, especially in East Asia, insects have represented the soul. In *Journey to the West*, Ch'eng-en Wu's mythological epic from late Medieval China, Old Monkey sometimes took the form of a fly to escape from demons or to elude detection.

Among the Montagnards of Vietnam, fireflies have traditionally been considered the spirits of departed heroes. In Japan and China, fireflies are the companions of impoverished scholars engaged in nocturnal study. Short poems, providing moments of illumination, which are written on fans or pieces of silk have been known as "fireflies."

The name of the demon Beelzebub, originally a Phoenician deity, literally means "Baal of the Flies," or "Lord of the Flies." In the Old Testament, he tempted King Ahaziah of Israel away from Yahweh (2 Kings 1:2-6), and later he was called the "prince of devils" by Matthew and Mark. In the Christian Middle Ages, demons were frequently depicted as flies, and so people often thought of insectivorous birds such as swallows as holy. There are several stories of devils taking the form of insects in order to enter peoples' bodies by flying into their mouths. According to a local chronicle, in 1559 a maiden near Joachimsthal in the Harz Mountains inadvertently swallowed an evil spirit, disguised as a fly, in her beer. The demon immediately possessed her and began to speak through her, though it was finally exorcised by the parish priest.

Before improvements in hygiene in the modern period, lice were often found in the hair and on the body of nearly everyone from king to peasant. Though a perpetual annoyance, they could also serve as a means of social bonding. To pick lice off a person was a service that might be performed by parents for children or servants for masters. It was even a ritual of courtship and love, performed by couples for one another. The presence of an inordinate number of lice might indicate either coarseness or, for ascetics, lack of worldly concern. Thus, Julian the Apostate, the austere Roman Emperor who attempted to revive paganism, once compared the lice running freely in his beard to wild beasts in a forest.

Fleas also tended to be thought of in a familiar and, at times, even in an affectionate way, though they were by far the most dangerous insects of the lot. It was not realized until the end of the nineteenth century, but fleas had been carriers of many diseases including bubonic plague. In the Renaissance, references to fleas became a humorous convention in poetic diction. Among the most famous is "The Flea" by John Donne, in which the author requests sexual favors from a woman by showing how their blood has mingled in the body of a flea:

> Oh stay, three lives in one flea spare,
> Where we almost, nay more than married are.
> This flea is you and I, and this
> Our marriage bed, and marriage temple is;
> Though parents grudge, and you, we are met,
> And cloistered in these living walls of jet.

When the young woman kills the flea, the speaker concludes:

> Just so much honor, when thou yield'st to me
> Will waste, as this flea's death took life from thee.

The words, however, certainly seem ironic in the perspective of today, when we know that fleas carried bubonic plague, which was entirely capable of wiping out entire villages in the early seventeenth century when this poem was written.

But insects, like rats, are now often put in the service of medicine. Apart from human beings and perhaps rodents, the *droposphila* fruit fly has become the most studied animal in the world. Scientists have found that the genetic code of the fruit fly is easy to manipulate and has many affinities with that of human beings. In the hope of correlating them with parts of the genome, all features of the creature's life from anatomy to courtship dances have been intricately observed. Nicholar Wade, a journalist for *The New York Times*, has observed that researchers who study fruit flies "are easily provoked into confessing that they think of people as large flies without wings."

HEDGEHOG

Not the phoenix, not the eagle, but the he'risson (hedgehog), very lowly, low down, close to the earth.

—Jacques Derrida, *Che cos' è la poesia?*

The Greek poet Archilochus wrote in the latter seventh century B. C. that, "The fox knows many tricks; the hedgehog knows only one. A great one." Everybody understood that Archilochus meant the animal's ability to elude predators by rolling itself into a ball, so that its quills would be facing in every direction. The broader meaning of the epigram, however, has puzzled readers for millennia. The poet had fought as a soldier, so perhaps he was thinking of a defensive military formation in which several soldiers with spears stand back to back in a circle. Perhaps the hedgehog may also have represented the poet's native Sparta, where people specialized in war, while the fox represented more cosmopolitan cities such as Athens. At any rate, Archilochus clearly considered the defense of the hedgehog to be at least the equal of the wiles of the fox. The twentieth century British philosopher Isaiah Berlin divided thinkers into foxes like Tolstoy, distinguished by breadth of understanding, and hedgehogs like Dostoyevsky, distinguished by their depth.

Such thinkers were doubtless also fascinated by the singularity of hedgehogs, which may appear a bit like miniature porcupines (to which they are not closely related) but resemble no other animal. They are small, nocturnal insectivores that burrow in the earth and, as already noted, are covered with spikes. They sleep beneath the ground for long periods when food is not plentiful then eventually reemerge, which made the ancient Egyptians associate the hedgehog with the renewal of life. The Egyptians, who constantly had to contend with the bites of snakes and the stings of scorpions, also admired the resistance of hedgehogs to poisons. They often carried amulets in the form of a hedgehog for protection against venomous creatures.

But people have feared as well as admired the powers of this diminutive animal. In Europe, hedgehogs were taken for companions of witches during the Early Modern Period. In China as well, the hedgehog has had a reputation for necromancy; people believed that it lay concealed near roads to cast spells on unsuspecting travelers.

Human beings are, like hedgehogs, unique among animals, and so the strangeness of hedgehogs can make these animals easy for people to identify with. Aristotle reported in his *Historia Animalia*, that hedgehogs copulate belly to belly like human beings, since the spikes make it impossible for the male to mount the female from behind. Aristotle, Pliny, and Aelian reported that hedgehogs can anticipate changes in the direction of wind; accordingly, they block and open entrances to their burrows. In the Byzantine Empire, people sometimes attempted to predict the weather by observing the burrows of hedgehogs.

In European fairy tales, the hero is very often the "simpleton," the child who at first seems too odd to participate in normal life. A fairy tale from the collection of the Brothers Grimm entitled "Hans my Hedgehog," began as a farmer complained, "I want to have a child, even if it's a hedgehog." Soon his wife gave birth to Hans, human to the waist and a hedgehog above. Hans' behavior was as strange as his appearance. He rode around on a rooster, played the bagpipes, and tended pigs in a forest. Though scorned and mistreated by people, Hans eventually managed to marry a beautiful princess and become fully human.

The hedgehog often appears as the proverbial "underdog" who manages to defeat a seemingly invincible opponent. In another tale from Grimm, the hedgehog challenges an arrogant hare to a race. After a few steps, the hedgehog slips beneath the ground. His wife, however, waits at the finish line, pretending to be her husband, and claims victory when the hare approaches. The tale concludes with a moral: "No person, no matter how superior he believes himself, should ever make fun of another, even if that other person is a hedgehog."

In Beatrix Potter's *Mrs. Twiggy-Winckle* [Tiggy-Winkle] (1905), the central character is a hedgehog that embodies values such as charity and hard work, which are romantically associated with a bucolic, pre-Industrial English village. Once famed for its formidable defense, the hedgehog has now come to symbolize the vulnerability of nature in a technological world. Crushed or wounded hedgehogs are a regrettably frequent sight on European roads. The Prickly Ball Farm in southeast England near Exeter has established a hospital and a network of volunteers to care for and rehabilitate these injured animals.

In an essay entitled "What is poetry?" (*Che cos' è la poesia?*), contemporary French philosopher Jacques Derrida compared a poem to a hedgehog that is thrown onto a street and curls up into a ball. In a similar manner, he maintained, a poem creates a self-contained world, which, however, exists amid terrible dangers. It may well be that Derrida, who himself is often accused of being solipsistic, identified with the proverbial hedgehog more than he cared to admit.

PIGEON

> *Pigeons in the grass, alas.*
>
> —GERTRUDE STEIN

Charles Darwin first theorized that pigeons were descended from rock doves, found along the Eurasian coasts, and this has recently been confirmed by genetic analysis. When people began to practice agriculture in Neolithic times, rock doves were attracted by the crops and extended their range, entering into a symbiotic relationship with humankind. They have been trained to carry messages from early times, and Greeks in the era of Homer used pigeons to announce the winners of the Olympic Games. Pigeons have been used to carry military communications, at least from ancient the era of ancient Greece and Rome up through World War I. In the sixteenth century, the Mughal Emperor Akbar kept an enormous colony of pigeons, which he bred largely for their appearance.

People have generally regarded the dove as sacred and the pigeon as utilitarian, but the two are somewhat conflated in Christian tradition, where the dove—sacralizing the pigeon's role as a courier— becomes a messenger from God. The dove of the Annunciation was at times portrayed as a white rock dove, with a very broad fanlike tale rather than the narrower tail of the turtle dove and related varieties. Pigeons had been regularly eaten in the ancient world, but their culinary use may have later been partly inhibited by religious symbolism. In the modern era, pigeon meat has rarely been a staple, but people have turned to it in times of scarcity and war.

The passenger pigeons of North America, once so numerous that they darkened the skies, were driven to extinction in the early twentieth century. Now they are remembered as a symbol of human rapacity and the lost bounty of the New World. Poet Wallace Stevens probably had the passenger pigeon at least partially in mind when he wrote "Sunday Morning" in 1915. It closed with the lines:

> And, in the isolation of the sky,
> At evening, casual flocks of pigeons make
> Ambiguous undulations as they sink,
> Downward to darkness, on extended wings.

But pigeons generally blend in so well with our urban environments that most people hardly even notice them. The few who do pay attention find much beauty in their enormous variety of patterns and tones, caused largely by the mixing of wild and feral birds. Pigeons thrive in cities because the facades of buildings resemble the stony landscapes of their original home.

Feeding pigeons and other birds as a recreation may be very old, and could go back to a sort of sacrifice. After World War II had ended and prosperity began to return, it became a tradition for tourists and their children in St. Mark's Square in Venice and Trafalgar Square in London. It was a symbolic expression of benevolence, as well as a celebration of abundance, but both cities banned the activity on hygienic grounds as the first decade of the twenty-first century drew to a close.

Today, people often refer to pigeons contemptuously as "rats with wings." There are small but devoted circles of pigeon fanciers, who race them and compete in shows. While lovers of many animals such as horses and cats tend to be female and aristocratic, pigeon enthusiasts are generally male and blue-collar. They identify with the toughness of these birds, which can survive easily on the most desolate coasts and in the roughest neighborhoods.

RAT AND MOUSE

> *Rats! They fought the dogs and killed the cats,*
> *And bit the babies in the cradles,*
> *And ate the cheeses out of the vats,*
> *And licked the soup from the cooks' own ladles,*
> *Split open the kegs of salted sprats,*
> *Made nests inside men's Sunday hats,*
> *And even spoiled the women's chats*
> *By drowning their speaking*
> *With shrieking and squeaking*
> *In fifty different sharps and flats.*

—ROBERT BROWNING, "The Pied Piper of Hamelin"

For the most part, rodents may be rivals and enemies of people, yet the two have a paradoxical intimacy, a bit like a married couple that cannot live in harmony yet find it impossible to separate. Rats and mice can adapt to a vast range of different environments, and they are quite capable of living without human beings. Nevertheless, they thrive particularly well in urban settings, where human beings inadvertently provide them with vast quantities of food and enclosures for shelter. As hosts of the fleas that carry plague, rodents may have killed untold millions of human beings in the course of human history. Even today, all our technologies cannot prevent rats and mice from devouring almost a quarter of the grain grown for human consumption. In the West, rats often appear in nightmares, and they can inspire revulsion and terror. Nevertheless, their ability to survive frequently earns grudging respect and admiration from people. In the Orient, rats are associated above all with prosperity, since they gather wherever food is plentiful. A Japanese proverb goes, "Getting rich is to invite the rat."

Most of folklore up through at least the Renaissance distinguishes only loosely between rats and mice. In Greek and Latin, both animals were generally designated by the word "mus," which is the origin of our word "mouse." The word "rat" comes originally from the Vulgar Latin "rattus," a term which probably originated in the Middle Ages. Like other pairs of closely related animals, such as lions and tigers, people have polarized these rodents as opposites, so that in the West the mouse has become beloved while the rat has become despised. In ancient manuscripts, people usually tend to translate the word "mus" according the whether rodents in question seem large and aggressive, like rats, or small and passive, like mice.

It was not until the nineteenth century that new techniques of construction enabled people to make buildings "ratproof"; prior to that, rodents were found in every structure from the barn to the royal palace. This produced a sort of intimacy with rats and mice, which may have softened the anger at the damage that they did. Rodents surely spoiled many meals and even destroyed homes, so it is remarkable that they were not often demonized in the ancient world. People might see rats and mice only occasionally, but they could hear them all the time, especially when falling asleep at night. They could not help but wonder, often with a certain sympathy, what transpired in the secret society on the other side of those holes in the wall.

One early attempt to imagine this is the fable known as "The Town Mouse and the Country Mouse," inserted by the Roman poet Horace in his *Satire II*. A country mouse once entertained a city mouse in his humble hole, offering him a few scraps of bacon and remains of vegetables. The city mouse would hardly deign to touch such fare. He explained to his rural companion that, since life was short, he should make the most of it by spending his time amid more pleasant surroundings. A little while later, the country mouse accepted the city mouse's invitation to dinner. The host brought in course after course of fine dainties left over from a banquet the evening before. The guest was rejoicing in his good fortune, when, all of a sudden, somebody started banging on the doors, and the entire house trembled at the barking of two ferocious hounds. The terrified country mouse took his leave, saying he would rather live humbly in peace than risk his life for sumptuous delights. The fable, a classic expression of the contrast between the city slicker and the country bumpkin, has been constantly retold, often set in contemporary urban centers such as New York or London.

People are continually amazed at the ability of rodents to get food, no matter how carefully it seemed to be locked up. Up through the nineteenth century and even today, they have tried to explain this with countless anecdotes, in which admiration for the ingenuity of rodents almost always seems to cancel any resentment of them as pests. Many authors have described how one mouse or rat would lie on its back and hold an egg in its paws, in order to be dragged like a sled by colleagues. Others told how mice stood on the shoulders of one another to form a living ladder, in order to reach food on a table. Quite a few popular writers maintained that rodents had elaborate rituals and customs such as burying their dead.

But neither the affection for mice nor the admiration for ingenuity of rats could ever overcome the practical necessity of keeping the rodent population under control. The Egyptians sometimes depicted mice with affection, but they also kept mongooses and cats in their homes to catch rodents.

According to Herodotus, an Egyptian king named Sethos had once alienated the warrior class by claiming their ancestral lands. When the Assyrian Sennacherib invaded Egypt, the warriors refused to support him. The King, who was also a priest of the sun-god Ra, entered the inner sanctuary of the temple, prayed, and wept until he fell asleep. The god appeared to him in a dream and told him not to worry. He should gather whatever soldiers he could, even if they were only merchants or artisans, and go forth to face the enemy. The two armies were encamped opposite one another. On the night before the battle, a swarm of field mice entered the camp of the Assyrians. They devoured the bowstrings and quivers of the enemy, leaving them weaponless. A statue of Sethos was later erected in the temple of Ra. In his hand, the King held a mouse. The inscription read, "Look on me, and fear the gods."

Since the lion is a symbol of kingship, it seems possible that that story may be the ultimate origin of the Aesopian fable, "The Lion and the

*llustration to the fable of
"The City Mouse and
the Country Mouse" by
Richard Heighway, 1910*

Mouse," told the by the Roman freedman Phaedrus and many others. A lion had caught a mouse, which begged to be let go, saying he might some-day return the favor. The lion was so amused at the idea that so tiny a crea-ture could ever help the king of beasts that he magnanimously lifted his paw and spared the mouse. A while later, hunters caught the lion in a trap. The mouse passed by and, seeing his friend struggling haplessly, gnawed away the ropes and set the lion free. The eternal rivalry between cat and mouse became a favorite theme of storytellers from the fables of Aesop to Tom-and-Jerry cartoons in twentieth century America. In one popular fable from the Middle Ages, the mice met in council to decide what they might do about the cat. They agreed that the greatest danger from the cat lay in the silence of his approach. One mouse proposed that a bell be tied around the neck of the cat to warn them of his approach. The members of the council applauded until one old mouse got up and asked, "Who will bell the cat?"

C. S. Lewis used this motif in his novella *The Lion, the Witch, and the Wardrobe*, where mice gnaw through ropes binding the lion Aslan, who represents Christ, to a stone table. Japanese tell how the Medieval painter Sesshu was once tied up during his childhood as punishment for idling away his time with art. He drew pictures of rats by moving his feet in the sand, which were so vivid that they came to life and gnawed his bonds.

A modern rendering of this theme is Edgar Allen Poe's famous story, "The Pit and the Pendulum." Rats had tormented a man who has been tied up in a dungeon by the Inquisition, but they ultimately liberated him by gnawing away his bonds. Like many stories of liberation by rats or mice, this tale can be read as an allegory about the emancipation of the soul at death. Because of their preference for human dwellings, rodents have often been taken for the souls of the departed. Because of their association with the next world, they are often credited with clairvoyance. According to a popular superstition, rats know when a ship is about to sink, and their leaving a ship is a sign of impending doom. The phrase "like rats leaving a sinking ship" is a scornful condemnation of those who abandon a cause at the first sign of danger.

In another Greco-Roman fable traditionally attributed to Aesop, a farmer once noticed that a mountain was rumbling, rocks were tumbling down, and dust was coming from its summit. He decided that the mountain was in labor, and he called his companions to see what it might give birth to. As they gazed on in fear and wonder, a tiny mouse finally emerged and came running down the slope. The story may well have originally referred to the emergence of the soul from the body. Sometimes a rodent also represents the separable soul, which can run about while a person is in a trance or asleep. In the Walpurgis Night episode in the first part of Goethe's *Faust*, the protagonist dances with a young witch at a nocturnal revel, but he is horrified when a rat leaps out of her mouth and runs away. Today, however, people speak of how a "mountain labored and gave birth to a mouse" to describe disappointing results after a great effort.

This idea of rodents as the souls of human beings seems to underlie the mysterious tale of the Pied Piper of Hamelin, which was recorded in several versions during the Middle Ages. In 1284 the town of Hameln in Germany became infested with rats, and the village council hired a brightly dressed piper to get rid of them. He played a mysterious tune that compelled the rats to follow him until he led them into the Weser River to drown. The Piper disappeared for a while, but he returned on Saint John's day to demand the payment that had been promised to him. When the village refused to pay what the Piper wanted, he began to play his pipe again, and this time all of the village children followed him. A mountain opened up to receive the procession and closed up after it, so the children were never seen again. The Grimm Brothers made the story famous in their collection of German legends, and Robert Browning, Goethe, and others have retold it. Various scholars have traced the tale back to the bubonic plague, to the Children's Crusade of the Middle Ages, or to a German migration southwards to Bohemia. However that may be, the image of the Piper with the children or rats greatly resembles Medieval representations of Death leading the departed in a dance.

There is also at least a very strong association between rodents and the dead in the legend of Bishop Hatto of Mainz, Germany. There was a famine, but Hatto continued to dine in luxury, and refused requests from the poor and hungry to lower the prices on his ample store of grain. Finally, weary of hearing the starving people complain, he invited all who lacked bread to assemble in a huge barn. Then, instead of offering food, he set the barn on fire and burned his visitors to death, while scornfully comparing them to rats. Next morning, the Bishop rose and saw that rats had eaten his portrait. A servant informed him that rats had eaten everything in the granary. He looked out over his lands to see a huge army of rats descending on the palace. In terror, the Bishop fled to an island in the Rhine and locked himself up in a structure known today as the "Mouse Tower." The rats followed, gnawed through the door, and finally ate up the villain alive.

The perspective on rats in East Asia is far more unequivocally positive. A legend tells that when the Buddha was near death all the animals came to pay their last respects. The ox was leading the way, when the rat hitched a ride upon its back. As they reached the pavilion where Buddha lay, the rat jumped down, raced ahead, and arrived before the other animals. As a reward for its piety, the Buddha granted the rat the first position in the Chinese zodiac.

Daikoku, the Buddhist god of wealth is often depicted holding a large bale of rice, at which rats nibble. These rodents serve him as messengers. The amazing fertility of rodents makes them symbolic of the way money can increase through good business, though even Daikoku has sometimes had to guard his store from rats.

In the Middle Ages, rats were sometimes viewed as familiars of witches or forms in which sorcerers ran about at night. However, it was not until some centuries after the worst episodes of the bubonic plague that we start to see intense expressions of aversion and disgust for rats, as people gradually began to suspect their connection with disease.

The reputation of rats took a drastic turn for the worse at the end of the nineteenth century when the French missionary Paul Louis Simmond announced that bubonic plague had been caused by a bacillus that was found on fleas carried by rats. The disease may have been around since the advent of humankind, and one of the first likely references to it is in the Bible. In the early eleventh century BCE, the Philistines had defeated the Hebrews and taken the Ark of the Covenant. "The hand of Yahweh weighed heavily on the people of Ashdod (Philistines) and struck terror in them, afflicting them with tumors"(1 Samuel 5:6). Outbreaks of the plague gradually became more common and more severe with the growth of trade and the increasing density of population during the Roman Empire. The plague of Justinian in 531-532 CE killed tens of millions, depopulating entire towns, but the most terrible outbreak of all was in 1348-50, when bubonic plague destroyed over one third of the population of Europe. New research, carried out by epidemiologists in the early twenty first century, has, however, seriously placed in question whether rats were the major carriers of fleas with the deadly bacillus, which could also have been spread by other means including voles and direct human contact.

The plague had also sometimes been blamed on Jews, and thousands of them were burned alive in the Middle Ages in consequence. In the latter nineteenth and twentieth centuries, rats have often been used in anti-Semitic propaganda. Cartoonists made the proverbial "Jewish nose" appear like the snout of a rat. In the Nazi propaganda film "The Eternal Jew," directed by Fritz Hippler, the migrations of Jews were compared to the spread of rats across the world. The physician Hans Zinsser, doubtless thinking of the two world wars, has observed that the conflict between the brown rat, indigenous throughout Eurasia, and the black rat, brought to Western Europe on the boats of crusaders, was a very close equivalent to armed conflict among human beings. In George Orwell's novel *1984*, the fate most dreaded fate by the hero Winston is to be eaten by rats.

As the rat has been demonized, the mouse, as though in compensation, has generally grown more beloved. In 1923 Walt Disney, then a struggling entrepreneur, introduced the first animated film starring Mickey Mouse as "Steamboat Willie," a captain who raucously hooted and danced as he steered his ship. As Disney Studios grew into a giant corporation, Mickey became more subdued and, in the eyes of his critics, even bland. The Mickey Mouse Club was founded as part of the television show "Walt Disney Presents." It featured boys and girls wearing caps with large mouse ears who sang, danced, and had adventures.

Meanwhile, medicine has increasingly placed mice and rats in the service of humankind, as the subjects of medical experiments. In 1988 the first patent ever was issued for an animal other than microorganisms. The "onco-mouse" was genetically engineered to develop cancer so it may be used in research. Over twenty million rodents, perhaps far more, are killed in American laboratories every year, and the scale of this use is probably increasing. To see what treatment is right for an individual patient, doctors now use mice as personal avatars, by simulating the patient's condition as closely as possible in the laboratory animal. In some cases, bacteria from a human patient's stomach are transplanted into mice, so doctors can experiment with different treatments. Eventually, researchers hope that rodent avatars will even produce customized immune cells, which can be injected into an ailing man or woman. This may create an unprecedented intimacy between the patient and the laboratory animal, since the rodent dies for its human double, in a way that is too direct and immediate to ignore.

Common adder and ringed snake

13
From the Underground

THE FIRST FEW INCHES OF FOREST SOIL ARE WHERE BOTH THE DECAY OF organic matter and the regeneration of new life are most pronounced, and it is often hard to tell life from death. It is also a dwelling place of nature deities, such as Spider Woman in the mythology of the Navaho and Hopi Indians. In many highly formalized religions, especially the Abrahamic ones, the realm beneath the ground is Hell, where demons torment, cook, devour, and excrete sinners. For the fifteenth century CE alchemist Paracelsus, often considered "the father of modern medicine," creatures of the earth (also called "earth elementals") were "gnomes." Our ground is their air, and they can see and walk through it. They are very small yet basically human in form, but, like people, they can have deformed births, which might take all manner of strange shapes. Fanciful as this may be, it does in ways describe life in the topsoil, where beetles and centipedes bore through the ground as easily as we walk over it.

ANT

In the ant's house, the dew is a flood.

—Persian Proverb

A plague had destroyed the people of King Aeacus, and he begged Zeus to give him citizens numerous as the ants in a sacred tree. Zeus changed the ants into warriors. These were the Myrmidons, who later fought under Achilles. Ants resemble warriors in many ways: they march in columns; they show

Soldier ants on sentry duty.

(Illustration by Ludwig Becker)

unbounded courage. No matter how large the adversary, they still attack. No matter how many are killed, they do not surrender or retreat. An ant that is decapitated will continue to bite at its adversary. For their size, ants are also the strongest creatures in the world, able to carry objects many times larger than they are.

According to another myth, Zeus changed himself into an ant in order to make love to the maiden Eurymedusa in Thessaly. She gave birth to a child named Myrmidon, ancestor of the martial race. Myrmidons were not only efficient warriors, but they also prospered in peace. Like ants, they would diligently work the soil.

The ants have regular access to the mysterious depths of the earth, where metals and jewels are found. Herodotus told of ants in India that were larger than foxes. These ants threw up huge heaps of sand that contained gold as they burrowed in the ground. The Indians watched from a distance, then quickly heaped the sand into bags and rode away on camels. They had to rely on surprise in these raids, since the ants were extremely swift in pursuit.

Ants holding an assembly in a pear tree.

(Illustration by Ludwig Becker)

In the fable of the industrious ant and the feckless grasshopper, attributed to the half-legendary Aesop (see Grasshopper), the ants appear almost as ruthless in their diligence as in their wars. "Idler, go to the ant; ponder her ways and grow wise," said the Bible, in a passage traditionally attributed to Solomon (Proverbs 6:6). Around the world, the ant has long been proverbial for industry.

Creatures that live beneath the earth, near the world of the dead, are always frightening and mysterious. At festivals of the dead, Jains and certain Hindus feed the ants. West African tribes have traditionally believed that ants carry messages from the gods. In ancient Greece and Rome, ants sometimes appeared in prophetic dreams. When King Midas was a child, ants carried grains of corn to his lips as he slept, a sign that he would one day achieve enormous wealth.

According to Plutarch, when the Greek commander Cimon sacrificed a goat to the god Dionysus during a war with the Persians, ants swarmed around the animal's blood. They carried blood to Cimon and wiped it on his big toe, predicting his immanent death. Ants are still used in divination. To step on ants brings rain. A nest of ants near your door means you will grow rich.

La Fontaine's version of "The Grasshopper and the Ant."

(Illustration by J. J. Grndville, from Fables de la Fontaine, 1839)

For all their reputation for ruthlessness, ants in folklore often protect the weak and vulnerable In the story "Cupid and Psyche" by the first-century CE Roman author Lucius Apuleius (in his novel *The Golden Ass*), the young girl Psyche had fallen in love with Cupid. The goddess Venus, mother of Cupid, did not approve of their union. She captured Psyche, locked the maiden up with a huge heap of mixed grains, and then demanded that all be sorted by nightfall. The ants pitied Psyche and carried the different grains, one by one, to separate piles.

According to Cornish legend, ants were fairies, which had, over the centuries, grown ever smaller and are now about to disappear. Other legends made them the souls of unbaptized children, who were admitted neither to Heaven nor to Hell. All such tales reveal a kinship that people feel with ants. Part of the reason may be a similarity of their bodies to ours, with large heads and hips but slender waists. It may also be that their small size and consequent vulnerability elicits our sympathy.

Ants appear in many European tales as grateful animals. In one fable by La Fontaine, a dove uses a blade of grass to a rescue a drowning ant. Later, a hunter tries to shoot the dove. The ant bites the man on the heel and makes his arrow go astray.

In Aztec mythology, red ants once kept the seed of maize hidden in a mountain. The god Quetzalcoatl transformed himself into a black ant and stole the seed in order to bring food to humankind. As in so many European stories, grain formed a bond between ants and humanity. The Hopi Indians traditionally believe that the first human beings were ants.

In *Walden*, Thoreau reported going to a woodpile to find a battle between two varieties of ants, with the ground already "strewn with the dead and dying." In one camp were "red republicans, "in the other, "black imperialists." "On every side they are engaged in deadly combat . . . and human soldiers never fought so resolutely," Thoreau continued. A valiant Achilles among the red ants came to avenge some fallen comrade. He killed a black Hector, while the enemy cavalry swarmed over his limbs. The gentle hermit of Walden Pond wrote of this carnage with great excitement. Perhaps those who believe there are no more heroes today should spend more time around the anthills. The ants live in a world a bit like that of old romances, filled with monsters (that is, termites, spiders, woodpeckers, or human beings). Kingdoms with mysterious powers surround the anthill, and ants must constantly battle in order to survive.

Perhaps when kings and lords ruled most of the world, it was a bit easier to identify with ants. The anthill seemed to be the perfect authoritarian state, in which everyone unquestioningly accepted a role. As governments became more democratic, it was a bit harder to believe that such a society was either possible or desirable. Artists and writers tried to individualize ants, and that certainly has not been easy.

When people look closely at ants, there is no telling what they may find. Michel de Montaigne reported in "Apology for Raymond Sebond" that the philosopher Cleanthes once observed the negotiation between two different anthills. After some bargaining, the body of a dead ant was ransomed for a worm.

Rev. J. G. Wood, in his extremely popular collection of animal anecdotes entitled *Man and Beast* (1875), gave this report of a lady that had killed several ants: "After a while, one more (ant) came, discovered his dead companions and left. He returned with a host of others. Four ants were assigned to each corpse, two to carry it and two to walk behind. Changing tasks at times so as not to become weary, the ants eventually arrived at a sandy hill where they dug graves and buried the dead. About six ants, however, refused to help with the digging. These were set upon by the rest, executed and unceremoniously thrown into a mass grave."

Anthills are like metropolises or vast armies, and people can almost never distinguish between individuals. Perhaps, though, even ants can have their nonconformists. The Disney animated film *Antz* told of one ant named "Z," who was unable to work or dance in the same way as the others. Gradually, he moved the workers over to his style of thinking. He saved the colony from a flood and married the princess. Did they live happily ever after? They started having millions of children, but their anthill did not seem greatly changed. Perhaps the movie was about the loosening of social restraints in the West during the late sixties. "Revolutions may be fun," the message seemed to be, "but don't expect too much! Just like ants, we are ruled by our genes."

Nothing we can say about ants is ever quite right. They are not really "communist" or "authoritarian." They don't really even have "workers," "soldiers," "slaves," or "queens." As scientists discover more about these creatures, the experience of ants only becomes harder to imagine. French author Bernard Werber took up the challenge in his novel *The Empire of Ants*. The ants, he explained, communicate mostly by scent, using pheromones, so they are like one enormous mind spread across the globe. A young female ant, separated from her community, set out to found a new anthill. She explored the world of a yard, filled with beetles, termites and birds, until finally she became queen of one vast colony that was powerful enough to challenge human beings.

Well, ants really do not threaten us, but their society will almost certainly outlast humankind. Ants thrive everywhere, from the Brazilian rain forests to the tiny cracks between the pavements in New York. Though they may appear frail, ants are able to survive even nuclear tests. In the Renaissance, an ant was sometimes depicted devouring an elephant to show the changeable nature of all things.

BEETLE

The poor beetle that we tread upon
In corporal sufferance finds a pang as great
As when a giant dies.

—William Shakespeare, *Measure for Measure*

Beetles are often found near garbage, around excrement, and in dank, moist areas. People usually associate beetles with filth, squalor, and decay, but we can think of these environments as "compost" that will soon nurture life.. In the ancient Mediterranean region, people believed that living beings, particularly insects, sprang spontaneously from decomposing matter, and a few varieties of beetles came to be regarded as holy.

Foremost among these is the scarab beetle, also known as the "dung beetle," which has been represented in countless amulets, some of which were found wrapped in the cloth covering mummified corpses. The scarab lives off dung and is frequently seen near farms. It pushes a globe of dung to an underground burrow before consuming it, and to the ancient Egyptians the ball suggested the sun crossing the sky. The scarab was sacred to Ra, god of the sun. The god Khepera, associated with creation and immortality, was depicted with the head of a scarab beetle. The female scarab beetle also lays her eggs in a ball that she creates inside her hole. The Egyptians confused the two processes of feeding and reproduction in scarabs, thinking the eggs hatching from the ball were generated spontaneously from the earth.

The Egyptians held a number of other beetles sacred as well. One long slender click beetle (Agy*pus notodonta*) was sacred to Neith, a very ancient goddess associated with both fertility and war, and it was often depicted on amulets and hieroglyphs. In the later periods of Egyptian civilization, including the Ptolemaic and Roman eras, beetles were mummified and placed in miniature sarcophagi, in the expectation that they would enter the next world. According to Charbonneau-Lassay, the veneration of certain beetles eventually spread from Egypt to much of the area around the Mediterranean. It also spread south to sub-Saharan Africans, who revered the golden-tinted rose beetle up through at least the eighteenth century.

One brightly colored insect, usually orange with black spots, is known as the "ladybug" in the United States and the "ladybird" in Britain. "Lady" can refer to the Virgin Mary. In France, the insect is sometimes called "*poulette à Dieu*" or "chicken of God." These names and the regard with which the insect is viewed hint at a divine role in religions of antiquity, but this has never been fully explained. One theory is that the beetle was once sacred to Freya, the Scandinavian goddess of love. The association of the ladybug with the sun, however, indicates that its cult may be connected with that of the Egyptian scarab.

To have a ladybug alight on your clothes is considered a sign of good fortune, but you must never kill or injure the creature. You are to send it away with a rhyme, though you may also hurry it along with by blowing gently. In one common the British variant this goes:

> Lady Bird, Lady Bird,
> Fly away home,
> Your house is on fire,
> Your children will burn.

The verse may possibly refer to the ladybug's association with the sun, which could be partly due to the bright red on its back. Another theory is that the lines could refer to the burning of hop vines after harvest to clear the fields. Sometimes, if the request is made in additional verses, people have believed the insect will fly to one's sweetheart.

In traditional rural societies, the cycle of putrefaction and renewed life was evident in yearly routines such as fertilizing the land, but it is experienced less intimately in Western cultures today. The beetle, a form of "vermin," sometimes appears as a symbol of unredeemed filth. In the novella *The Metamorphosis* by the German-Jewish writer Franz Kafka (1915), a traveling salesman named Gregor Samsa, subject to degrading demands from both family and work, woke up one day to find he had been changed into a giant insect. Many readers take it for a cockroach, which is now perhaps most intimately associated with squalor and decay. The creature in Kafka's story is indeterminate, but the description of the insect lying on its hard back with legs kicking helplessly in the air suggests a beetle. By accepting his martyrdom, to die neglected and be thrown out in the trash, without resentment, Gregor becomes a Christ-like sort of figure.

SCORPION

> *Do not forget Yahweh, your God who brought you out of the land of Egypt, out of the house of slavery: who guided you through this vast and dreadful wilderness, a land of fiery serpents, scorpions, thirst . . . "*
>
> —DEUTERONOMY 8:15

Since the scorpion is found in cracks, crevices, holes, and enclosed places, it is often associated with chthonic deities. Much like a snake, it can often strike unexpectedly. Its sting can be very painful and sometimes deadly. The scorpion can be a symbol of evil, but also an instrument of divine retribution.

This dual symbolism is already found in ancient Egypt. The god Set, evil brother of Osiris, attacked the infant Horus in the form of a scorpion, no doubt a familiar sort of experience to the ancient Egyptians since children were particularly vulnerable to the sting of this animal. The divine infant was saved by the medicine of Thoth, god of knowledge. Selket, a goddess of marriage, fertility, and the underworld could, however, also use the power of the scorpion to fight demonic powers. She was often depicted with the body of a scorpion and the head of a woman, sometimes also as a scorpion holding an ankh, the Egyptian cross. In addition, she was painted as a woman with a scorpion on her head or in her hand, an image that the Romans eventually began to use as an allegorical representation of the African continent.

Selket, a scorpion-goddess of ancient Egypt.

Scorpion-man from an inlaid harp found in excavations of Ur, circa 2600 B.C.

The Hittite and Babylonian equivalent of Selket is the goddess Ishara, who is associated with love and with motherhood. In her honor, the scorpion first became a constellation of the zodiac among the Semitic peoples of Mesopotamia. Scorpion-men frequently appear in the art and literature of Mesopotamia. They usually have the heads of human beings, the wings and talons of birds, a serpent for a penis, and the tail of a scorpion. Sometimes they are associated with Tiamat, the sinister mother-goddess, but they are also attendants of the sun god Shamash and guard his realm against demons. In the Epic of Gilgamesh, a scorpion-man and scorpion-woman guard the mountain where the sun rises.

According to a Greco-Roman legend, the great hunter Orion boasted that he would kill all animals. On hearing that, the earth-mother Gaia summoned a scorpion, which bit his heel and killed him. Asclepius, the divine physician, restored Orion to life, but Zeus would not accept this interference in the process of life and death, and so he sent a thunderbolt that killed Orion a second time. Placed in the zodiac, the scorpion and Orion now represent death and life, whose eternal battle is enacted as the sun moves between the two constellations.

In China the scorpion was one of the five venomous animals, but images of it could help to keep demons or illness at bay. In Zoroastrianism, the scorpion was a creation of Ahriman, the power of darkness. In Mithraic religion, however, the scorpion became one of the animal companions of Mithras, as he sacrificed a great bull to regenerate the world.

In Christianity, the scorpion was often a symbol of the devil, waiting in ambush for unsuspecting travelers. Jesus told his followers, "Yes, I have given you power to tread underfoot serpents and scorpions . . ." (Luke 10:19). In the late Middle Ages and Renaissance, the tempter in Eden was sometimes depicted not as a serpent but as a creature with the face of a woman and the tail of a scorpion, an image which goes back to Selket. Some misogynistic moralists have compared the scorpion's habit of hiding in holes to a woman who would hide perfidious intent behind a beautiful face.

SNAKE AND LIZARD

Now the serpent was more subtle than any beast of the field which the Lord God had made. —GENESIS 3:1

Snakes can have dozens of young at a time, and so they are often symbols of fertility. They resemble vegetation, especially roots, in their form and often in the green and brown of their skins. The undulating form of a snake also suggests a river. A point of muscular tension passes through the body of a snake and drives the animal forward, like a moment in time moving along a continuum of days and years. Like time itself, a snake seems to progress while remaining still. In addition, the body of a snake also resembles those marks with a stylus, brush, or pen that make up our letters. Ornamental alphabets of the ancient Celts and others were often made up of intertwined serpents. It could even be that the tracks of a snake in sand helped to inspire the invention of the alphabet. The manner in which snakes curl up in a ball has made people associate them with the sun.

According to one legend, Sakyamuni, who later became the Buddha, was once walking beside a cliff when he looked down and saw a great dragon renowned for wisdom. The seeker asked many questions, and the dragon answered all of them correctly. Finally, Sakyamuni asked the meaning of life and death. The dragon replied that he would only answer when his hunger had been stilled. Sakyamuni promised his body as food, and the dragon revealed the ultimate truth. Then Sakyamuni hurled himself into the open jaws of the dragon, which suddenly changed into a lotus flower that carried him back to the precipice. The snake (in this case, a dragon) is an eternal mediator between opposites: good and evil, creation and destruction, female and male, earth and air, water and fire, love and fear.

Since the snake does not have exposed sexual organs, it is very hard to tell the male snakes from the female ones. Serpents often represent a primeval androgynous state before the separation of male and female. In the ancient world, however, serpents were associated with a vast number of goddesses, including the Greek Athena, the Mesopotamian Ishtar, the Egyptian Buto, and many others. Serpentine divinities include the Babylonian Tiamat, a primeval goddess from whose blood the world was created. The pharaohs of ancient Egypt would wear the ureaus on their heads, a protective image of the goddess Wadjet in the form of a cobra, leaning back and ready to strike. For the aboriginal people along the northern coast of Australia, the rainbow serpent is the most important of spirits, and they see traces of it in the undulating rivers, valleys, rocks, and caves of their landscape.

As people turned more exclusively to patriarchal deities, there was a massive revolt against the cult of the snake. This is why serpents are so often destructive in the mythologies from very early urban civilizations. Egyptians believed that the serpent Apep would try to devour the boat of the sun god Ra as he sailed it through the earth every night. Serpents have been killed by just about every major god or hero of the ancient world, and many in Medieval times as well. The Babylonian Marduk killed the serpent goddess Tiamat, and Zeus killed the primeval serpent Typhon. Apollo, the son of Zeus, killed the serpent Python to gain the shrine at Delphi, formerly sacred to the goddess Gaia. As an infant in his crib, Hercules strangled two serpents to death. Cadmus, a legendary founder of Greek civilization, killed a serpent and then planted its teeth, from which sprang warriors; the ancestors of the

Drawing of a sea serpent reportedly sighted in 1743.

noble families of Thebes. Sigurd, the Norse hero, killed the dragon Fafnir. Saint George—patron of England, Russia, and Venice—killed a dragon, while Saint Patrick drove the snakes out of Ireland. Even today in Texas, some communities have annual "rattlesnake roundups." People festively collect rattlesnakes, tease them, and finally kill them for food.

Images of the snake are often similar in cultures that appear to have little or no contact with one another. In Aztec mythology, the earth mother Coatlicue was a female serpent in a primordial sea. She is today especially well-known because of a monumental Aztec statue of her, dated from the fifteenth century CE. The statute, located in the National Museum of Anthropology in Mexico City, is one of the most terrifying images ever made. She has a double head; two rattlesnakes gazing at one another. She wears a skirt woven entirely of serpents, and a necklace of human hearts and skulls. According to Aztec myth, she gave birth, without prior intercourse, to many deities including Huitzilopochtli, the sun god, and Quetzalcoatl, the god of wisdom. In one story, her elder daughter, Coyolxauhqui was enraged at their birth, and, together with Coactlicue's previous children, turned against her mother. Huitzilopochtli, emerging from the womb fully armed, then cut off Coyolxauhqui's head and threw it into the sky, where it became the moon, while the siblings allied with her became stars. Coatlicue shares many features with Tiamat, including a terrifying visage, a close association with snakes, and an ability to conceive children without a partner. It is possible that they could both go back to a single mythology that arose before human beings had spread from Eurasia to the Americas.

After expelling Adam and Eve from Eden for eating from the tree of knowledge, Yahweh placed a curse on the serpent, which has ever since crept upon the ground. But just as the Biblical Yahweh—who can often be wrathful, jealous, and arbitrary—does not seem unequivocally good, so the serpent of Eden does not always appear fully evil. In the Middle Ages and Renaissance, the serpent of Eden was often painted with a human head, usually that of a woman: In *Paradise Lost*, Milton describes the serpent thus:

> Woman to the waist, and fair,
> But ended foul in many a scaly fold
> Voluminous and vast, a Serpent arme'd
> With Mortal sting.

At times, the serpent is a mirror image of Eve. Even in paintings in which the head of the serpent is bestial, the serpent and Eve often seem to be exchanging meaningful glances, while Adam simply looks confused. Eve and the serpent share a feminine wisdom. The serpent of Eden is also identified with Lilith, the first wife of Adam, who was also a Sumerian goddess-demon.

The large, intense eyes of the snake appear very mysterious. Pliny the Elder and countless subsequent writers have reported that snakes can hypnotize and even kill with a simple gaze. The basilisk, a serpent with a crown and wings, reportedly had this ability, as did the rattlesnake in the United States. Many authors from journalists and novelists to serious natural scientists reported that snakes could draw birds out of the sky by looking upward, and they sometimes even worked their powers of fascination on human beings.

The serpent has frequently been rehabilitated and even deified, especially by the Gnostics and the alchemists. Serpents are ancient symbols of healing. In the Mesopotamian Epic of Gilgamesh, the serpent steals the plant of immortality then sheds its skin and lives forever. Ancient physicians from Greece to China realized that venom extracted from certain serpents may be used to cure ailments such as paralysis.

Serpents are associated with the Greek healer Asclepius, who once raised a man from the dead. The ancient Greek physicians or "Asclepiads" had so much confidence in the healing powers of the serpent that they would sometimes place snakes in the beds of those with high fevers. They may have served as a sort of placebo, and the coolness of the serpents flesh could have convinced the sufferer that he was recovering. The Caduceus, a wand with two serpents entwined around it was carried by Hygeia, daughter of Asclepius, and by the god Hermes. Today it remains a symbol of the medical profession.

The alchemists saw the serpent as an animal that joined all of the four elements out of which the cosmos was formed. Of all animals, serpents are most intimately associated with the earth. This further associates them with fire, since that element escapes from the earth in volcanoes. The red tongue of many snakes, ending in a fork and flickering in and out, also suggests flame. Dragons, especially in European traditions, often breathe fire. Furthermore, serpents are also frequently found in water, and their rhythmic motion suggests waves. Many dragons and other serpentine figures are depicted with wings.

Among the most popular images among the alchemists was the ouroboros, a snake with its tail in its mouth, a symbol of primal unity. This goes back at least to the Egyptian Book of the Dead, written around 1,500 BC, and it was later taken over by the esoteric religions of Greece. An analogous figure is the serpent Midgard of Norse mythology, which is coiled around the earth. The Chinese used the "V" shaped fangs of a serpent to symbolize the essence of life, while upside down this represented the spirits of deceased ancestors.

From the point of view of folklore, lizards may generally be regarded as snakes, even though most (not all) lizards have legs. Because these animals are often found in lying in the desert sun, they have sometimes been associated with contemplative ecstasy. Pliny the Elder had reported, with some skepticism, that the salamander, a black and yellow lizard found in most of southern Europe, would quench flames with the coldness of his body. Paracelsus, an influential alchemist and physician of the Renaissance, believed that the salamander was a being of pure fire. The salamander sitting inside a furnace became a symbol of esoteric knowledge. It was compared to the three young Hebrews who were thrown into a fiery furnace by the king of Babylon but were not harmed by the flames (Daniel 3:22-97), and to Christ descending into Hell.

Even the Hebrews, who reacted so vehemently against the archaic cult of the serpent, have occasionally attributed godlike, supernatural powers to this animal. In the book of Exodus, Moses and Aaron were demanding of Pharaoh that the people of Israel be released from bondage. To demonstrate the power of his god, Aaron threw his staff down in front of Pharaoh and his court. It immediately turned into a serpent. At the direction of Pharaoh, the court magicians also tossed down their staffs, which became serpents as well, but Aaron's serpent swallowed all the others. Later, on the journey to Canaan, the Hebrews were stricken by a plague of fiery serpents. Moses directed the people of Israel to erect a bronze serpent on a standard, and all those who look upon it were saved from death (Numbers 21:4-9). Were this not sanctioned by scripture, it would probably have seemed to the Jews like sorcery and idolatry. Among the most extravagant dragons of all was that which did battle with Saint Michel in the Biblical book of Revelations. It had seven heads, each bearing a crown and ten horns, and it swept a third of the stars from the sky with its tail.

A positive view of the snake has also frequently been preserved in folk culture. During the wanderings of Aeneas, after the fall of Troy, his father Enchases had died. Landing on the coast of Sicily, he began the funeral rites by pouring out wine, milk, and blood of sacrificial animals. Then he cast flowers upon the funeral mound and began his oration. He had barely begun to speak when, in the words of Virgil's *Aeneid*:

> . . . from the depths of mound and shrine a snake
> Came huge and undulant with seven coils,
> Enveloping the barrow peaceably
> And gliding among the altars. Azure
> Flecks mottled his back; a dappled sheen
> Of gold set all his scales ablaze, as when

A rainbow on the clouds facing the sun
Throws out a thousand colors.
            Aeneas paused,
Amazed and silent, while deliberately
The snake's long column wound among the bowls
And polished cups, browsing the festal dishes,
And, from the altars where he fed, again
Slid harmlessly to earth below the tomb.

The snake was the spirit of his father, whom Aeneas would later visit in Hades. Romans would sometimes feed snakes at household altars.

The Scythians who lived by the Black Sea and were known for their fierceness traced their ancestry to the daughter of the Dnieper River, who was a woman above the waist but whose body ended in a serpent's tail. Not only the ancient Romans but many also other peoples—for example, Australian Aborigines—have believed that ancestors return in the form of snakes. Legendary Zulu kings would also sometimes be reincarnated as powerful serpents.

Despite, or because, they are not easily distinguished by gender, snakes appear highly sexual, and there are many tales of serpentine paramours. One tale from the Hindu-Persian *Panchatantra* tells how a Brahman and his wife longed for children but had been unable to conceive. One day a voice in the temple promised the Brahman a son that would surpass all others in both appearance and character. A short time later, his wife did indeed become pregnant, but she gave birth not to a human being but to a snake. Her friends advised her to have the monster killed, but she insisted on raising the snake as her child, keeping him in a large box, bathing him regularly and feeding him with fine delicacies. At her urging, the Brahman even arranged for the snake to marry a beautiful girl, the daughter of a friend. The girl, who had a strong sense of duty, accepted the marriage and took over the care of the reptile. One day, a strange voice called to her in her chamber. At first she thought a strange man had broken in, but it was her husband, who had climbed out of the snakeskin and taken on human form. In the morning, the Brahman burned the snakeskin, so his son would not

Frogs surround a snake in this illustration by Kawanabe Kyôsai.

be transformed again, and then proudly introduced the young couple to all the neighbors.

Both snakes and dragons are designated by the same word, "draco," in Latin. We can generally regard dragons as snakes, just as did zoologists of the Middle Ages and Renaissance. Edward Topsell wrote in *History of Four-Footed Beasts, Serpent and Insects* (1657), "There be some Dragons which have wings and no feet, some again have neither feet nor wings, but are only distinguished from the common sort of Serpents by the combe growing upon their heads and the beard under their cheeks." The variety and range of dragons vastly exceed those of any other mythic animal. Dragons often combine features of many animals, such as the wings of bats or horns of stags, but these are set upon serpentine forms. Just as it mediates between the elements, the snake seems to combine features of all creatures in its incarnation as the dragon.

In very archaic times, the serpent was almost universally revered in China. Into the twentieth century CE, several temples in southern China have followed the tradition of keeping sacred serpents that are offered wine and eggs on the altar. One popular legend tells that when Buddha was meditating prior to attaining enlightenment, a snake approached and wound itself around him seven times, and then guarded him for seven days from the other animals of the forest. Chinese culture gradually began to distinguish more sharply between the snake and the dragon, usually to the detriment of the snake, and they became opposites in the Chinese zodiac.

The Chinese dragon, known as "lung," is among the most colorful and extravagant composites. When first born, it appears as a simple serpent. Over thousands of years of life, it acquires the head of a camel, the scales of a carp, the horns of a deer, the eyes of a hare, the tusks of a boar, and the ears of an ox. It also has four short legs with enormous claws, a mouth with long teeth, and a flowing mane running down its back. A combination of fire and steam issues from its nostrils to form the clouds, and so it controls the weather. By tradition, conception in the year of the dragon brings special blessings, and so Chinese communities always experience an increase of births around that time.

As the modern period began, people increasingly thought of the snake as masculine. The traditional eroticism of the snake was originally considered primarily a feminine attribute, and only later a male one. This view was sanctioned by Freudian psychology, where people have usually interpreted the snake as phallic. In James Joyce's *Portrait of the Artist as a Young Man*, the hero, Stephen Daedalus, calls his sexual organ "the serpent, the most subtle beast of the field."

Legend usually locates fantastic beasts on the frontier of human exploration, and the serpent is a good example. With the expansion of maritime trade at the end of the Middle Ages, the Great Sea Serpent was second in importance only to the mermaid as a figure in the lore of mariners. Sightings of serpentine creatures were reported everywhere from Loch Ness in Scotland to the coasts of the New World, often attested by many people who had reputations for good judgment and sobriety. These were identified with many mythological creatures that ranged from the Norse Midgard Serpent to the Biblical Leviathan. While the descriptions differed in their details, they generally described the serpent as extremely long and as moving with an undulating motion. On August 21, 1936, for example, newspapers reported a monster seen by several Newfoundland fishermen that was at least 200 feet long, has "eyes as big as an enamel saucepan," snorted blue vapor from its nostrils, and stirred up such waves that "for days no boat dared venture out to sea."

The symbolism of the snake has changed far less fundamentally than that of other animals such as the dog or horse. It seems to surface whenever people contemplate origins, whether of humanity, of life, or even of the universe itself. Today, the DNA code that directs the development of the embryo is sometimes called "the cosmic serpent."

WORM

I am a czar—a slave; I am a worm—a god.

—Gavil Derzhavin

Through most of history, people generally have not distinguished sharply between earthworms, eels, and snakes. The word "worm" or "wurm" was sometimes used, in English and other Germanic languages, to designate a snake or dragon up through much of the nineteenth century. The earthworm, however, has never had the satanic glamour of serpents. With facial features that are difficult to see, earthworms are unusually hard to distinguish from one another. In consequence, they are seldom thought of as individuals, and, even in mythology and folklore, they rarely speak.

Worms have been associated since ancient times with moist, fertile earth. In Antiquity, worms were believed to be generated spontaneously from dirt, in part because they would appear after the rain. This primeval character was shown in a Mesopotamian incantation entitled "The Worm and the Toothache," written in the second millennium BCE:

> After Anu (god of the sky) had created heaven,
> Heaven had created the earth,
> The earth had created the rivers,
> The rivers had created the canals,
> The canals had created the marsh,
> And the marsh had created the worm—

The worm went, weeping before Shamash (god of the sun),
His tears flowing before Ea (god of deep water):
What wilt thou give me for my food?

The gods first offered the worm a fig, but it refused and asked to be placed in the gums of a person.

In literature, a worm is usually a symbol of humble status. A psalmist, for example, has written, "Yet here am I, now more worm than man/ scorn of mankind, jest of the people . . ." (Psalms 22:6). The book of Isaiah concluded with an eschatological prophecy that God would make the heavens and earth anew. In the final words, God told the prophet how all people will come to worship:

And on their way out they will see
the corpses of men
who have rebelled against me.
Their worms will not die
nor their fire go out;
they will be loathsome to all mankind.

Worms eat corpses, so to live on as part of a worm was the destiny of all who failed to transcend the body.

In the late Middle Ages and Renaissance, Europeans became increasingly obsessed with physical decay, and worms symbolized the corruption of the body after death. Worms were often painted crawling in and out of a human skeleton. The following lines from Shakespeare's *Hamlet* showed a noteworthy awareness of ecological cycles:

HAMLET: A man may fish with the worm that hath eat of a
king, and eat of the fish that hath fed of the worm.
KING: What dost thou mean by this?
HAMLET: Nothing but to show how a king may go a progress
through the guts of a beggar.

The worm, the most humble of animals, was always triumphant in the end, showing the ultimate equality of all living things.

Just as serpents are traditionally associated with moral corruption, worms are often connected with physical decay. Their habit of burrowing inside a fruit or flower has often impressed people as a botanical equivalent of demonic possession, and it can seem vaguely sexual. This is apparent in "The sick Rose," written by William Blake in the last decade of the nineteenth century:

> O Rose, thou art sick.
> The invisible worm
> That flies in the night
> In the howling storm
> Has found out thy bed
> Of crimson joy
> And his dark secret love
> Does thy life destroy.

But the connection of worms with the earth could be restful as well as disturbing.

The very last study written by Charles Darwin before his death was entitled *The Formation of Vegetable Mold Through the Action of Earthworms.* Some scientists have regretted that the founder of evolutionary theory should have turned to such a specialized theme in the end, but the choice reflected his desire for bucolic peace. Darwin wished to show that earthworms, traditionally the lowest of creatures, were an essential part of the natural cycle. They improved the fertility of the soil and so are essential for the continued prosperity of other creatures, including humankind. With the growing ecological awareness in the latter twentieth century, earthworms increasingly symbolize the ultimate unity of all life. This is even apparent in the field of physics, scholars of which refer to discontinuities in time and space as "wormholes."

14
BY THE SEASHORE

The seashore has long been a place of wonder, where one can still find colorful stones, shells, bits of coral, smooth fragments of glass, and the relics of all sorts of strange and wonderful events beneath the waves. A few creatures such as turtles and seals are almost (though not quite) equally at home in two realms, earth and water; others, such as gulls, add the third realm of air. Often, the seashore symbolizes the intersection of time and eternity. In folklore, it is the abode of shape-shifters such as seals and swans, which cast off their skins to become human, and may then marry men or women.

The seal and dolphin might have been included here, but, since they also inhabit the depths of the sea, they will be found in Chapter 7 on "Mermaid's Companions."

SEAGULL AND ALBATROSS

At length did cross an Albatross,
Thorough the fog it came;
As if it had been a Christian soul,
We hailed it in God's name.

—SAMUEL TAYLOR COLERIDGE, *The Rime of the Ancient Mariner*

Mariners traditionally observe seabirds closely, since the behavior of these creatures can tell them about subtle changes in the weather or their distance from land. Interpreting the flight and calls of avians, however, is a fairly intuitive art in which it is still often hard to distinguish reasonable calculation from superstition. The flight of birds can legitimately tell sailors a lot about things like the strength of winds or the distance from land, but it can also inspire daydreams.

Seagulls are remarkable flyers, able to ascend to great heights or hover on the wind. The ability of seagulls and related birds to glide on currents of the wind while remaining almost motionless makes them resemble a cross, the symbol of Christ, when seen from below. These birds follow ships, as though drawn by some kinship with human beings. Their calls are loud and confident yet also plaintive. Gulls and other sea birds can also easily appear to be spirits, especially when their white feathers are seen at night against a dark sky. Sailors from at least Medieval times to the present have believed that gulls, albatrosses, and stormy petrels were the souls of those drowned at sea. Three gulls flying overhead together are an omen of death. To kill an albatross or gull brings bad luck, and the fisherman should immediately release any bird caught in his net. Should a seagull fly against the window of the sailor's house while he is away, it is a sign that the master is in danger.

Ovid wrote that once, when the goddess Venus, the Greek Aphrodite, had appeared on the fields before Troy, the impetuous Greek hero Diomed wounded her. Later, as he returned home after victory, his boat was tossed about by terrible storms. The sailors knew this was the vengeance of Venus. One hotheaded crewman named Acmon heaped his scorn on the goddess and challenged her to do her worst. His companions rebuked him, and he tried to answer. The words did not come, for his voice had grown thin; his mouth had become a beak, and his arms were covered with feathers. He and his companions had been turned into white birds, gulls and their various relatives. In *The Golden Ass* by Apuleius, a gull would fly about the world to bring back news to Venus.

Ovid also wrote in his *Metamorphoses* that when Ceyx, king of Thrace, was preparing to journey by sea to consult an oracle, his wife Alcyone was seized by foreboding. She pleaded with him to remain. She reminded him of the shards of ships washed up on the shore. When Ceyx insisted on leaving, she begged to accompany him, for, she said, should the ship go down in a storm, at least they would lie together. After a long pause, Ceyx refused, promising to be home again within two months' time. The premonitions of Alcyone proved true, and the ship went down in a terrible storm. Juno, the goddess of marriage, sent a dream of Ceyx to Alcyone. Her husband appeared naked and pale, and water flowed through his hair and beard. She woke and ran to sea by the light of dawn. In the distance, she could see a body. As it floated towards her, she slowly recognized her husband. As Alcyone ran to him, her feet skimmed over the surface of the sea; her arms changed into wings, her nose into a beak. Ceyx rose from the sea, and the couple became a pair of birds. Since that time, the winds and sea stay calm for seven days in winter as Alcyone nests on the surface of the waters. Ovid does not identify the birds, but tradition makes the wife a halcyon, a half legendary bird mentioned by many authors from Homer on, which scholars believe was a mythologized kingfisher. The name "Ceyx" means "tern," but Ovid was far more interested in vivid tales than in ornithology, and he may have thought of both husband and wife simply as any seabirds.

The albatross is reportedly able to predict the weather. In Japan the albatross is a servant of the sea god and, therefore, auspicious. In the west, however, an albatross following a ship has been considered a herald of storms. In Samuel Taylor Coleridge's poem *The Rime of the Ancient Mariner*, the narrator killed an albatross that accompanied the ship, apparently on a perverse impulse but perhaps also to evade a tempest. His ship was then stopped by a terrible calm, and the crew draped the dead albatross about his shoulders as a cross. Only when he had learned to love his fellow creatures did the bird fall from his neck, and the ship begin to move.

SWALLOW

True hope is swift and flies with swallow's wings . . .

^WILLIAM SHAKESPEARE, *Henry V*

The swooping, gliding motion of a swallow as it catches insects in flight, together with its incessant calls, can at times appear melancholy. In much of the world, people think of the swallow as a joyful bird, since its presence announces the coming of spring. Since the swallow often makes its nest in crannies of buildings, it has always been on intimate terms with human beings. In France and other parts of Europe, rural people have often believed that the nest of a swallow would protect their homes. For Romans, swallows represented the penates or household spirits. Sometimes swallows were believed to be spirits of dead children, and people were not permitted to kill them. Swallows are almost perpetually in the air, so much so that Medieval people thought they did not have feet. Partly for that reason, they have always been considered very spiritual.

Swallows were often depicted on Egyptian mummies, suggesting they were already associated with resurrection. In an ancient love poem from Egypt, quoted by Dorothea Arnold, a young woman returns from her lover:

> The voice of the swallow is speaking.
> It says:
> Day breaks, what is your path?
> (The girl answers) Don't little bird!
> Are you scolding me?
> I found my lover on his bed,
> And my heart was sweet to excess.

The poem shows a remarkable resemblance to the "songs of dawn" by Medieval minnesingers and troubadours, in which lovers would often be awakened by a lark or other bird.

Plutarch related that, when Isis had retrieved the coffin of Osiris, she flew about his body and lamented in the form of a swallow. In Italian legend, when Mary stood before the Cross, swallows saw her distress and came to comfort her. They swooped down, coming ever closer, until finally they began to touch her in passing with their feathers. As they turned, the tears of Mary landed on their breasts and turned them from black to white. Swallows have frequently been portrayed darting about Christ as he lay on the cross. According to Swedish legend, they fanned him with their wings. For

*The swallow and spider
compete for a fly.*

*(Illustratiion by
J. J. Grandville, from* Fables
de la Fontaine, *1839)*

devout Christians, the appearance of swallows traditionally suggests the resurrection. Insects, especially flies, have often been associated with the Devil, and swallows, like agents of God, would relentlessly pursue bugs and snatch them up in flight.

People have always watched for swallows as a sign that winter was at an end. A fable recorded by the Hellenized Roman Barbius told of a young man who saw a swallow, thought that spring was coming, and decided to wager his winter clothes on a throw of the dice. After he had lost, a snowstorm came, and he soon saw the swallow lying dead from the cold. "Poor creature," he said, "you fooled both yourself and me." This tale may be the origin of the famous saying "One swallow does not make a spring."

In part because the trajectories of their flights resembled those of bats, many Europeans have believed swallows hibernated in caves. In a famous letter of August 4, 1767, the British pastor and naturalist Gilbert White made one of his rare mistakes. After a large fragment of a chalk cliff had fallen during one winter in Sussex, England, many swallows were reportedly found dead in the debris. Though a bit skeptical of the account, White thought the dead swallows might have been in hibernation. About six and a half years later, White observed that the first swallows of the year were usually seen near ponds and that they disappeared in the event of frost. He suggested that the birds lay dormant through the winter under water. A few naturalists of the seventeenth and eighteenth centuries even believed that swallows sojourned on the moon. In Muslim countries swallows have traditionally been considered holy, on the ground that they make a yearly pilgrimage to Mecca. Their migratory paths were not actually mapped out until the nineteenth century.

DUCK, GOOSE, AND SWAN

> *But now they drift on the still water,*
> *Mysterious, beautiful;*
> *Among what rushes will they build,*
> *By what lake's edge or pool*
> *Delight men's eyes when I awake some day*
> *To find they have flown away?*
>
> —W. B. YEATS, "The Wild Swans at Coole"

Swans, geese, and ducks are aquatic birds, which appear more comfortable on water or in the air than on land. They tend, however, to congregate along the shore, since that is where food is most plentiful. If a stranger approaches a pond, they may all rise in unison. Because they often act in concert, they seem to have a sort of solidarity across lines of species. In Sanskrit, a single word, *hansa*, was used to designate all three varieties of birds. Nevertheless, their personalities in folklore have become very distinct. The swan is poetic, solitary, and often tragic in myth and legend, perhaps because of its white plumage and extraordinary grace. The duck is generally not alone but found together with its mate and children. The folkloric goose is more gregarious and earthy, and it is often part of a noisy flock.

These birds had considerable religious significance in prehistoric times. Throughout Eurasia and the Near East, there are many myths of the world being hatched from a cosmic egg. The Egyptians, for example, believed that Ra, the god of the sun, had emerged from the egg of a goose. Many ancient figurines have been found combining the features of water birds, such as elongated necks and beaks, with those of human females. Several scholars such as Marija Gimbutas have speculated about the worship of a bird-goddess in Neolithic times.

The Aspares of archaic Hindu mythology are water nymphs that transform themselves into waterfowl, and are perhaps early versions of swan maidens, which are such important figures of Eurasian folklore. There are countless stories, especially in Scandinavia and other lands in the far north of Eurasia, of water birds which become women and marry into human society, only to leave their husbands, resume their old form and fly away. Jacob Grimm and many other scholars have connected the swan maidens with the Valkyries, the warrior women of Norse mythology who appear in the sky to lead those slain on the battlefield into Valhalla.

Zeus took the form of a swan to seduce the maiden Leda, who then gave birth to the heroic twins Castor and Polydeuces. Leda herself, however, may once have been a swan deity. According to some versions of the legend, she laid two eggs. From one egg hatched the twins and from the other Helen of Troy. According to yet another version of the tale, given by Apollodorus, Nemesis, the goddess of fate, tried to escape the amorous attentions of Zeus by turning into a goose, but the god changed into a swan and raped her. Nemesis laid an egg, which was found by a shepherd in the woods and brought to Leda, who placed it in a chest. Helen eventually hatched from the egg, and Leda raised her as a daughter.

The swan was sacred to the Greco-Roman sun god Apollo, and it appeared on Greek coins as early as the third century BC. In early Greek religion, people sometimes believed that swans drew the chariot of Apollo across the sky each day, though horses later replaced them. The Greek philosopher Plato was known as the "swan of Apollo." Socrates, according to tradition, had once dreamed that a fledgling swan had flown to him from an altar consecrated to love, rested for a while on his knees, then flew away singing beautifully. As he finished relating the dream, Plato was introduced

Traditional Chinese motif of ducks as symbol of conjugal love.

to him, and Socrates knew immediately that this boy had been the swan. Just before his death, Plato dreamed that he was a swan flying from tree to tree as people tried in vain to catch him. Simmias, a former companion of Socrates, interpreted the dream to mean that many would try to grasp his spirit yet no interpretation would capture the full meaning of his words.

A very widespread legend—found in Aelian, Pliny the Elder and many other writers of the ancient world—is that swans sing a supernaturally beautiful song as they die. In the late Middle Ages, the swan pierced with an arrow and singing as it swam became a symbol of the house of Lusignan in France. Swans were especially linked with death in Irish mythology, in which they were the form taken by lovers to cross the boundary between earthly existence and the Otherworld. In one myth, Oenghus, the god of love, fell in love with a maiden named Caer. She changed into a swan on the feast of Samhain (November 1), and flew to join him. He also became a swan, and together they flew around a village three times and sang the people to sleep. The term "swan song" is still used to describe a final, valedictory achievement.

In another popular Irish legend, when Lir had been deposed as ruler of Ireland, he married Aev, and together they had three boys and one girl. A short time afterwards, however, Aev died. Lir remarried, this time to Aoife,

Aev's sister. The stepmother could not bear the way everyone admired the four children, since they were not her own. One day, as the children were bathing, Aoife took a magic wand and transformed them into swans. The girl Fionnula, eldest of the siblings, begged Aoife to give them back human form. Aoife would not relent, but at last she agreed to place a limit on the enchantment. She let the children retain human voices, and ordained that they must wander the desolate islands for nine hundred years before they would again become human. When the allotted time was almost up, they passed Saint Mackevig, a follower of Saint Patrick, on the remote island of Innis Gluaire. Mackevig heard the children singing as they swam by, marveled at their voices and wished to win them for his choir. After much searching, Saint Mackevig found the children of Lir, and led them into his chapel. When the bell rang for communion, the curse ended. In place of the swans were four ancient human beings. Saint Mackevig had just time to bless them and baptize them before they died.

The story of Lohengrin, the swan knight, is a sort of swan maiden story with the genders reversed. It was included in the epic *Parzifal* and many other Middle High German manuscripts, and the Grimm brothers eventually included several versions in their collection of German legends. Though often embellished with colorful details, the essential story is as follows. Duchess Else of Brabant was pressured to marry, but she had rejected her suitors. One day the knight Lohengrin sailed down the Rhine River in a boat drawn by a swan. Else received him, and the two soon fell in love, married, and had children. Lohengrin defended the kingdom valiantly, but he warned his wife never to ask about his origins. One day, thinking that her children should know about their father, she inquired about Lohengrin's family in a moment of forgetfulness. The knight then returned to his swan-drawn boat and sailed away, never to be seen again.

The pristine white plumage was, for centuries, a fundamental characteristic of swans in the European imagination, but black swans were discovered by English settlers in Australia around the end of the eighteenth century. Whiteness had always been identified with goodness and purity, and so very the very existence of black swans challenged the bird's traditional symbolism. Economist Nassim Nicholas Taleb has coined the expression "black swan," used to refer to any event that is not predictable but has an enormous impact. In his words, ". . . almost no discovery, no technologies of note, came from design and planning—they were just black swans."

Perhaps people were too intimidated by the poetic magnificence of swans to ever domesticate them very successfully. The cackling of geese, however, has been heard constantly on farmyards since ancient times. Since they become excited and honk noisily at any disturbance, they alert people to any threats. For this reason, they were thought of as protectors of the home. Geese were sacred to Hera, the Greek goddess of marriage, and to her Roman counterpart Juno. They were often kept in temples. When the Gauls invaded Rome in 390 and were scaling the Palatine Hill by night, the dogs remained silent, while the geese alerted the defenders and saved the city. According to Aelian, the event was commemorated yearly in Rome by a celebration in which a dog was sacrificed but a goose was paraded in a litter.

There are many tales of a goose that laid golden eggs, the best known of which comes from the fifth-century Roman fabulist Avianus. A farmer had such a magical bird, but he became impatient waiting every day for golden eggs. Finally, he decided to kill the goose, expecting to obtain the entire treasure all at once. Cutting it open, he found, to his distress, absolutely nothing inside. The moral is to be thankful for what you have and not to demand more.

People have distinguished sharply, however, between domestic geese and their wild counterparts, whose migratory habits remained unknown until modern times. In 1187, Giraldus Cambrensis, who explored the coasts of Ireland, stated that wild geese hatched from shells that clung to driftwood. Later authors reported that they grew on trees near the edge of the sea, and they were called "barnacle geese" or "tree geese." The famous botanist John Gerard wrote in his 1597 *Herbal* that he had come across several of these shells on an old, rotten tree, and taken them apart to find avian embryos in various states of development. These reports enabled people of the late Middle Ages and Renaissance to classify wild geese as fish rather than as meat, which meant they could be eaten on Fridays and during Lent.

Male authors from the Renaissance through the Victorian period tended to view women as being either utterly wild or completely domestic, and they had much the same attitude towards geese. Unlike swans or ducks, geese have always been thought of as feminine, in part because their incessant chatter made misogynistic men think of gossipy females. Charles Perrault gave *Tales of Mother Goose* (1697) as an alternative title of his famous collection of fairy tales. The frontispiece to the first edition showed an old woman spinning as she told tales to children. Since then people have debated the identity of Mother Goose, if indeed she were an actual person at all. However that may be, the designation "Mother Goose" soon came to refer to an archetypal teller of stories, a bit like Aesop in the ancient world. From at least the latter part of the eighteenth century, English and American nursery rhymes have been known as rhymes of Mother Goose. This apocryphal author has been depicted as a goose, at times wearing a bonnet like an old-fashioned nursemaid or housekeeper, or even as riding a huge, flying goose.

All water birds are monogamous. While swans and geese marry human beings in folklore, ducks appear content with their own kind. This is in part because, unlike swans and geese, male and female ducks are usually very distinct from one another in appearance. Ducks are beloved in the Orient for the variety and splendor of their plumage. Mandarin ducks are Chinese symbols of conjugal fidelity, and to kill them brings bad fortune. Lafcadio Hearn has retold a Japanese story of a hunter named Sonjo, who came upon a mandarin duck couple in the rushes, killed the male, cooked him, and ate him. That night, he dreamed that he saw a beautiful woman weeping bitterly. She noticed Sonjo, reproached him for killing her husband, and told him to go again to the rushes. The next morning, the female duck swam straight towards him, tore open her breast with her beak, and then died before his eyes. Sonjo was so shaken that he gave up hunting and became a monk.

In the West, ducks have become affectionate symbols of the modern bourgeoisie, who value domestic peace more than poetry or heroism. Hans Christian Andersen contrasted traditional and modern values, to the detriment of the latter, in his famous tale "The Ugly Duckling." One egg of a mother duck is slow in hatching; the chick that finally comes out is oddly proportioned and unusually large. It is not accepted by the other ducks, eventually rejected even by its own mother, and forced to wander about alone. Finally, the lonely bird flies up to join a flock of glorious white swans, fully expecting them to attack him; instead, they immediately welcome him to their flock. Looking at its reflection in the water, the bird finally realizes that it had been hatched from the egg of the swan, which had been laid among the ducks. The theme, a common one in the Romantic Movement of the early nineteenth century, was the suffering of the poet in the prosaic world of the middle class.

By the latter nineteenth century, swans were featured less in literature than in the highly stylized media of opera and ballet such as Richard Wagner's "Lohengrin" and Pyotr Ilyich Tchaikovsky's "Swan Lake," where they evoked a past of wonder and heroism. In Henrik Ibsen's drama "The Wild Duck," a wounded, wild duck, rescued by a young woman and confined to a barnyard, symbolizes the frustrations of domestic life.

Though the swan was favored in heraldry, the duck is far more prominent in popular culture of the modern age. Rubber ducks for the bathtub are among the most beloved of children's toys; wooden ducks are used by hunters as decoys and to decorate the mantelpiece. The most popular duck of all is Donald, who has been the subject of innumerable cartoons and comic books since the 1930's. He shows all the neurosis and insecurity of the middle classes. He is perpetually jealous of his companion Mickey Mouse, and is forever getting into trouble. All of this makes him easy to identify with in our relatively unheroic age, and, besides, he is usually pretty successful in the end. In Europe, he now far surpasses Mickey in popularity.

TURTLE AND TORTOISE

I am related to stones
The slow accretion of moss where dirt is wedged . . .

—ANTHONY HECHT, "Giant Tortoise"

Tortoises are generally larger than turtles and spend more time on land, but people did not even make a rough distinction between the two until the sixteenth century. The word "turtle" is still sometimes used as a general term for both, and that is how we will use it here. The folkloric reputation of the turtle as a primeval creature has, in some respects, found surprising confirmation by scientists. They have existed for about 230 million years, and individuals of some species can live over two centuries.

The wrinkled features of turtles suggest age, while their silence can give the impression of wisdom. Though not very fast even in the water, turtles have great strength and stamina. The shell of a turtle often represents the cosmos, with a dome for the sky and a flat surface beneath for the earth. When the turtle emerges from a shell, it can appear as new creation, while withdrawal into the carapace can seem like reversion to the beginning of the world.

Mythologies throughout the world have associated turtles with the primordial waters out of which the earth was formed. Turtles have often been used as symbols of fertility, since they are constantly seen copulating in ponds in spring. Furthermore, the head of a turtle emerging from the shell can suggest a phallic symbol. On the other hand, the resemblance of the carapace to a womb has moved people in some cultures such as the Chinese to think of turtles as primarily feminine.

Oppian, writing in Greek during the second century CE, believed that female turtles were always unwilling sexual partners and had to be raped by the males in order for the species to reproduce. Several modern observers have shared this impression, but it is hard to tell the feelings of turtles during copulation or at any other time. The apparent detachment with which turtles copulate helped make them symbols of chastity in the Christian Middle Ages.

In Greek mythology, the god Hermes, as a mischievous infant, invented the first musical instrument, the lyre, from a carapace. He saw a mountain tortoise grazing in front of a cave, and, perhaps thinking of the echoes in a cavern, killed the animal and strung the guts of sacrificed bulls across the shell. He later gave the lyre to the god Apollo in payment for stolen cattle. Perhaps in lost versions of the story, a nymph was transformed into a tortoise, so that she might escape the amorous advances of the god, much as Syringa was transformed into reeds when fleeing Pan. The lyre itself seems feminine, and playing it can easily suggest a sublimated sexuality. This would explain why it seemed inappropriate for Hermes to keep the instrument as his emblem.

According to legend, the *I Ching*, a Chinese system of divination, began in the early third millennium BCE as the fabled Emperor Fu Hsi was walking along the Yellow River and saw a turtle emerge from the waters. The sage ruler saw eight trigrams on the back of the turtle—that is, eight groups of three lines, either solid or broken, each. He interpreted these signs as patterns of cosmic energy that might be used to divine the future. Historians believe the *I Ching* may have actually been developed from a practice of divination that used the cracks in a tortoise shell exposed to intense heat.

A symbol carried on banners by the imperial army in China was a serpent wound around a turtle, though this image has been variously interpreted. Sometimes it has been taken as a struggle between the might of the serpent and the indestructibility of the turtle, in which neither party can be victorious. In another view, the serpent was male (or "yang"), while the turtle was female (or "yin"), and the two were engaged in copulation. It is not unusual, however, in East Asian cultures to have opposites such as strife and harmony represented by a single symbol.

In one very ancient story from Japan, known in many versions, a young man named Urashima was fishing all alone on the wide sea for three days yet had managed to catch nothing, when he felt a weight in his net and hauled up a multicolored turtle. As he lay down to sleep, the turtle suddenly changed into a beautiful young woman. She explained that she lived in heaven as a star of the Pleiades (or, in some retellings, dwelt beneath the sea) and had fallen in love with him. Urashima ascended with her to a heavenly mansion where they were married and lived together in happiness for three years. Then, for all the joys of his new life, Urashima began to miss his parents and begged his wife to allow him to visit them once more. She very reluctantly agreed, and gave him a box as a parting gift, telling Urashima to grip it firmly if he wanted to return, but to never open the lid. On returning home, Urashima found that the village was changed and his parents were long dead. He had failed to realize that a year in heaven was a century on earth. In panic, Urashima seized the box, thoughtlessly opened it, and instantly turned into a very old man, for the box contained all of the years he had spent in the celestial kingdom.

The second avatar of the Hindu Vishnu was as the turtle Kurma, who served as the base of the mountain Mandara, where it churned an ocean of milk for the deities, thus producing the nectar of immortality. According to some Hindu traditions, the world rests on the back of an elephant, which, in turn, is standing on a turtle. The idea that the world rests on a turtle's back is also found in the mythologies of several Native American tribes, especially those of the Eastern woodlands. According to a tale of the Huron Indians, variants of which have been recorded in many other tribes, the goddess Aataentsic, who lived in the clouds, split the tree of life with her ax. Part of the tree fell through a hole in the sky; Aataentsic, fearing that all life might perish, jumped down after it. At that time there was no earth but only water. When Turtle looked up and saw Aataentsic falling, he directed Beaver, Muskrat, Mink, and Otter to dive down and bring up earth from the bottom of the ocean. The animals placed the earth on Turtle's back to provide a cushion, so Aataentsic settled down and the tree took root. The Lenape and other tribes also have a myth that the earth was once covered by a great flood and human beings sought shelter on the back of a turtle.

In Africa, the turtle is a trickster figure, yet, unlike other tricksters such as the Native American coyote, he is virtually never impetuous. The slow movements of the turtle suggest caution, and its wrinkled face suggests the wisdom that comes with age. Other tricksters often become victims of their own cleverness and pride, but the turtle is prudent and almost always victorious. In one widely told West African story, which has also been recorded in the United States, a lion captured and tried to devour a turtle. The captive told the king of beasts, "Uncle, if you are wondering

Brer Terrapin is having a tug of war with Brer Bear, but the clever turtle ties his end of the rope to an underwater root in this Uncle Remus tale by Joel Chandler Harris.

(Illustration by A. B. Frost)

how to soften my shield and make it good for eating, just please put me to soak in the river." The lion obliged, and the turtle immediately swam away to hide in the mud.

This turtle deity was carried from Africa to the New World with the slave trade. When the slaves had to keep their practices and beliefs secret from their masters, they could draw inspiration from the silence of the turtle, as the humorous tale of "The Talking Turtle," recorded in many variants from Africa and the Caribbean to the United States, illustrates. One version from Alabama began as a slave was walking along one morning and saw a turtle by the edge of a pond. "Good morning, turtle," he said, being full of good spirits. "Good morning . . ." replied the turtle. As soon as he could recover enough composure, the man stammered that turtles cannot really talk." "You talk too much," said the turtle, and with that the creature slid into the pond. The slave ran home, and then returned with his master, to find the turtle once again in the sunlight on the edge of the pond. "Good morning, turtle," said the slave. The turtle did not reply. "Good morning," repeated the slave, a bit more insistently, but there was still no answer. "Liar!" shouted the master, and he beat the slave terribly. Later the slave returned to the pond and, finding the turtle in his accustomed place, began to reproach it for not having returned his greeting. The turtle responded, "Well, that's what I say about you Negroes; you talk too much anyhow."

In *Uncle Remus*, the collection of African-American folktales by Joel Chandler Harris, Turtle, or "Brer Terrapin," is the cleverest of the animals, able to consistently outwit even Brer Rabbit himself. Brer Terrapin lacks the cruelty of the other beasts, and its presence softens note of the nihilism in the stories. One tale began as the animals were at a picnic, and, while the females prepared the food, the male animals began to boast. Brer Rabbit said he was the swiftest, while Brer Bear claimed to be the strongest, and so on. Brer Terrapin listened calmly until all were finished, reminded the company of how the tortoise had won a race against the hare in a famous fable of Aesop, and then challenged Brer Bear to a test of strength. Brer Bear took one end of a rope, while Brer Terrapin took the other and slid into the pond. Brer Bear tried to pull Brer Terrapin out of the water, but the rope would not budge. Brer Terrapin had tricked his adversary by tying the rope to a root. When the pulling finally stopped, Brer Terrapin undid the knot and slid out of the water in triumph. The turtle has become a beloved figure in books for children, from Lewis Carroll to Walt Kelly and Dr. Seuss.

Butterflies

15
LOST SOULS

THE ANCIENT EGYPTIANS DEPICTED THE SOUL OR *BA*, AS A SMALL body with wings. This is usually taken as a bird, perhaps an owl, but in some pictures it seems less a small avian than a large insect. The Greeks and Romans pictured the soul as a butterfly or moth. The Christian gospels tell of the resurrection of the body, and make no mention of a soul. That disembodied self is originally a pagan concept, which entered Christianity through the work of Plato and other Greek and Roman philosophers. For alchemists of the Early Modern Period, the soul was a "homunculus," a diminutive human figure. Descartes believed that this little man was located in the pineal gland, right in the center of the human head, and gazed out through the eyes. Whatever its metaphysical status, we can understand the "soul" as a relatively ethereal aspect of a person or animal. The animals that can, at least at moments, represent the soul are small, and generally have a fluttering, spectral sort of flight or motion.

BUTTERFLY AND MOTH

> *I did not know then whether I was then a man dreaming I was a butterfly, or whether I am now a butterfly dreaming I am a man.*
>
> —CHUANG TZU

The idea of a butterfly or moth as the soul is a remarkable example of the universality of animal symbolism, since it is found in traditional cultures of every continent. The custom of scattering flowers at funerals is very ancient, and these attract butterflies, which appear to have emerged from a corpse. A butterfly or moth will hover for a time in one place or fly in a fleeting, hesitant manner, suggesting a soul that is reluctant to move on to the next world.

The transformation of a caterpillar into a butterfly provides the ultimate model for our ideas of death, burial, and resurrection. This imagery is still implicit in Christianity when people speak of being "born again." The chrysalis of a butterfly may have even inspired the splendor of many coffins from antiquity. Many cocoons are very finely woven, some with threads that are golden or silver in color.

The Greek word *psyche* means soul, but it can also designate a butterfly or moth. The Latin word *anima* has the same dual meaning. Several carved gemstones from ancient Greece depicted a butterfly hovering over a human skull. Late Roman artifacts often depicted Prometheus making humankind while Minerva stood nearby holding aloft a butterfly, representing the soul. A story inserted in the first-century novel *The Golden Ass* by the Roman-Egyptian author Lucius Apuleius tells of a young girl named Psyche who was given to Cupid, the god of love, in marriage, and contemporary illustrations often showed Psyche with the butterfly wings. Western painters have at times depicted the soul as a small human figure with such wings, an image also used to represent fairies.

In lands around the Eastern shores of the Pacific Ocean, the idea that the soul of a person will return in the form of a butterfly and hover around the grave of the body is widespread. In Indonesia and Burma, people have traditionally considered that a butterfly entering your house is likely to be the spirit of a deceased relative or a friend. On the island of Java, it is traditionally believed that, during sleep, the soul may fly out of the body in the form of a butterfly; you should never kill a butterfly, since a sleeping person might then die as well.

The Chinese sage Chaung-Tzu—a disciple of Lao-Tzu, the founder of Taoism—claimed that he fluttered about as a butterfly at night. On waking, he would continue to feel the motion of wings in his shoulders, and he was

Their close association with flowers helps to make butterflies a symbol of both fecundity and transience.

unsure of whether he was truly a butterfly or a man. Lao-Tzu explained to him that, "Formerly you were a white butterfly which . . . should have been immortalized, but one day you stole some peaches and flowers . . . The guardian of the garden slew you, and that is how you came to be reincarnated."

The way certain butterflies perform a courting dance—each partner moving off in various directions yet always coming back to the other—has made these insects symbols of conjugal love, especially in Japan. Lafacadio Hearn collected a Japanese story of an old man named Takahama, who was nearing death. A nephew was sitting at his bedside, when a white butterfly flew in. It hovered for a while and perched near his head, when his nephew tried to brush it away. The butterfly danced around strangely, and then flew down the corridor. Surmising that this was not an ordinary insect, the nephew followed until the butterfly reached a gravestone and disappeared. Approaching to examine the grave, he found the name "Akiko." On returning, he found Takahama dead. When the boy told his mother, she was not in the least surprised. Akiko, she explained, was a young girl that Takahama had planned to marry, but she had died of consumption at the age of eighteen. For the rest of his life, Takahama had remained faithful to her memory and visited her grave every day. The nephew then realized that the soul of Akiko had come in the form of a butterfly to accompany the spirit of his uncle to the next world.

The soul of a beloved also takes the form of an insect, probably a butterfly, in the ancient Irish saga of "The Wooing of Étaín." The god Mider fell in love with a mortal named Étaín, but the goddess Fuamnach, his first wife, struck the young woman with a rowan wand and transformed her into a puddle. As the water dried, it became a worm, which then changed into a "scarlet fly." "Its eyes shone like precious stones in the dark, and its color and fragrance was enough to sate hunger and quench thirst in any man; moreover, a sprinkling of the drops it shed from its wings could cure every sickness . . ." The insect accompanied Mider as he traveled and watched over him as he slept, until Fuamnach sent a fierce gale to blow it away. After being pursued constantly by the goddess for centuries, the insect was finally carried by wind into the goblet of a chieftain's wife, who swallowed it with her drink, and then gave birth to Étaín, giving her, once again, the form of a girl. Mider searched for her for a thousand years, but, when he finally found her, Étaín had become the wife of Echu, the king of Ireland. Echu refused to relinquish Étaín, but Mider defeated Echu in a board game, winning the privilege of embracing Étaín in the center of the King's hall. As they put their arms around one another, the lovers turned into swans, and flew away through the skylight.

As the pace of modern life has become increasingly frantic, people have come to admire the leisurely flight of the butterfly. As W. B. Yeats put it in his poem "Tom O'Roughley":

> 'Though logic-choppers rule the town,
> And every man and maid and boy
> Has marked a distant object down,
> An aimless joy is a pure joy,'
> Or so did Tom O'Roughley say
> That saw the surges running by,
> 'And wisdom is a butterfly,
> And not a gloomy bird of prey.

The way a butterfly moves from flower to flower has also been decried as lack of commitment, and Yeats, in the same poem, calls it "zig-zag wantonness." Today many ecologists regard butterflies as a keystone species, and they will count butterflies per acre in an attempt to determine the health of an ecosystem, perhaps in a manner not altogether different from that of diviners in the ancient world.

The distinction between butterflies and moths is not fully recognized by scientists, but, in folk culture, it is very simple—moths are nocturnal while butterflies are diurnal. Furthermore, butterflies have many dazzlingly bright patterns of color, while moths are usually shades of white and brown. When homes were lighted by candles or lanterns at night, people were particularly fascinated by those moths which would fly toward the fire, even when that meant they would expire in a sudden blaze. This image was often used by Sufi mystics to describe an ecstatic union with God. In one of his most famous poems, "Blissful Longing" (*Selige Sehnsucht*), Johann Wolfgang von Goethe used this motif as a symbol of the desire of the soul for transcendence. It tells of a moth drawn to a flame and ends with these words:

> And till you have stood this test:
> "Die, and come to birth!"
> You remain a sorry guest
> On this gloomy earth.

While many find the poem beautiful, some critics have been troubled by its romantic celebration of death.

James Thurber satirizes this longing for oblivion, countering it with a little romanticism of his own, in his fable "The Moth and the Star." A young moth conceived the idea of flying to a star, but his father said that was crazy, advising his son to focus on street lights, like any sensible young moth, instead. Ignoring parental advice, the moth kept trying to reach the star, and, though unsuccessful, lived out a full and contented life, while his moth, father, brothers, and sisters were all burned to death. The moral: "Who flies afar from the sphere of our sorrow is here today and here tomorrow."

A compassionate view of a moth's apotheosis in flame was expressed by the early twentieth century British author Virginia Woolf in her essay

"The Moth." She told of watching a moth dance about by day as its motions became gradually fainter. Many times she gave the moth up for dead, only to see it flutter once again. Finally, when the tiny body relaxed and then grew stiff, she felt awed by both the power of death and the courageous resistance of the spirit against so formidable an antagonist.

ENGLISH ROBIN AND WREN

Call for the robin redbreast and the wren,
Since o'er shady groves they hover,
And with leaves and flowers do cover
The friendless bodies of unburied men.

—JOHN WEBSTER, *The White Devil*

The connection between the English robin and the wren in folklore is so close that they have often been mistaken for the male and female of a single species. In Britain people say, "The robin and wren are God's cock and hen." The wren, which is grayish russet-brown, is associated with Christmas, while the robin, which is more brightly colored, is associated with Easter. According to a French legend, the wren fetched fire from Heaven, singeing its wings, and then passed the brand to the robin, which burned its breast.

In *Lark Rise to Candleford*, her account of growing up in rural England during the nineteenth century, Flora Thompson tells how boys would regularly take eggs from the nests of birds for food, decoration, and sport. The wren and robin, together with only a very few other birds, were spared because, to quote a saying:

The robins and the wrens
Be God Almighty's friends.

But the wren has been viewed with more ambivalence than its little cousin.

The wren is known throughout Europe as the "king of birds." In Latin, it is known as "regulus," in Greek as "basiliskos," and in Old German as "kunigli," all of which mean "monarch." The Roman author Suetonius claimed, in his history of the deified Julius Caesar, that the assassination of the dictator was foretold by the fate of a wren. Shortly before the dictator was killed, a wren, pursued by raptors and bearing a sprig of laurel, flew into the Hall of Pompey, named after Caesar's former rival, where it was overtaken and torn to pieces. But, if the wren is a monarch, it is one without finery, courtiers, or weapons. It rules the animals in something like the way the soul may be said to govern the body.

The status of the wren as king may seem paradoxical, since people usually associate greatness with size. As might be expected, veneration of the wren generally appealed to common people rather than to elites, which is probably why the cult has been far more prominent in folklore than in literary culture. Nevertheless, archeologist Edmund Gordon believed that the wren was celebrated in an animal proverb, from the early second millennium, from Sumer, the oldest urban culture. When the elephant boasted about his size, another animal, possibly the wren, answered him by saying, "But I, in my own small way, was created just as you . . ."

The wren seems to proclaim this uniqueness with a remarkably loud, melodious song, which will usually capture the attention of a person strolling through a meadow long before the bird itself is seen. The best-known version of the story explaining how the wren became king of birds is by the Brothers Grimm. The birds gathered to select a king and decided that the crown would be awarded to whichever could fly the highest. The eagle flew far above all the other birds and descended to claim his prize. The wren suddenly piped up that the crown was his. He had lain unnoticed on the back of the eagle until the large bird began to descend, whereupon the wren took off and flew so high that he could see God himself. Other birds objected that the contest had been won through trickery and so the result was invalid. The assembly decided that the crown could go to whoever burrowed most deeply in the earth. While other birds such as the rooster and duck began to dig, the wren simply found a mouse hole and descended. Though birds continued to object, the wren, to this day, slips in and out of hedges and proclaims, "I am king."

The story shows both the reverence and the ambivalence with which people have traditionally regarded the wren. In much of Europe, especially the British Isles, the wren is never killed throughout most of the year but is ceremoniously hunted on Saint Stephen's Day, December 26. The body of the slain wren is paraded through the streets on a pole in a colorful procession, to the accompaniment of music and song. Many legends are invoked to explain the ceremony. One held that the call of a wren betrayed Saint Stephen to his persecutors. According to an Irish legend, the Gaelic warriors were once planning a surprise attack on the troops of Oliver Cromwell, but wrens alerted the enemy by beating on drums.

Still another legend, this one from the Isle of Man, relates that there once lived a mermaid whose beautiful form and sweet song were irresistible to men. She would gradually lure them into ever-deeper waters until they finally drowned. Finally, a knight came along who was able to withstand her enchantment. He was about to kill the mermaid, but she escaped by taking the form of a wren. From that time on she was compelled to assume the form of a wren one day every year. People would kill the birds without mercy on that day, in hope of finally putting an end to the sorceress, and the feathers of the wrens were kept for one year as a charm against shipwreck. The fact that there are so many explanations for this ceremonial hunt suggests that it goes back very far, perhaps even to Neolithic times, and its original significance has been lost.

In western France, a similar ceremony was sometimes performed with the robin on Candlemas Day. The body of a robin was pierced with the branch of a hazel tree, which was then set aflame. The robin of folklore, as well as the wren, has been intimately connected with death and resurrection. Most of the legends about the robin center on the bright red color of its breast. According to one tale, the robin tried to pluck the thorns from the crown of Christ, whereupon blood fell on its breast. According to a legend from the Hebrides, the fire that warmed the Holy Family in the stable was about to go out, so the robin fanned the flames with its wings, thus burning its breast.

The mysterious English children's rhyme "Who did kill Cock Robin?" which provides an archetype for later murder mysteries, begins:

> Who did kill Cock Robin?
> I, said the Sparrow,
> with my bow and arrow,
> I killed Cock Robin.
>
> Who did see him die?
> I, said the Fly,
> with my little eye,
> I saw him die.
>
> Who did catch his blood?
> I, said the Fish,
> with my little dish,
> I caught his blood.

It concludes:

> All the birds of the air
> Fell a-sighing and a-sobbing,
> When they heard the bell toll
> For poor Cock Robin.

Linguistic analysis suggests that the poem may, in some form, go back to the fourteenth century CE or earlier. The widespread mourning indicates that, in this case, the robin, rather than the wren, is probably the king of birds. The earliest known version of the poem goes back to the mid-eighteenth century, when it was used as an allegorical allusion to the intrigues surrounding the abdication of Robert Walpole, the first Prime Minister of Britain.

The robin traditionally attended to the bodies of those who had been denied a proper funeral. In the anonymous British ballad "The Children in the Wood," an abandoned boy and girl perished in a deep forest:

> No burial these pretty babes
> Of any man receives,
> Till robin-red-breast painfully
> Did cover them with leaves.

But, for all the associations with martyrdom and death, the robin, like the wren, was generally thought of with joy as a herald of the spring. In the children's classic *The Secret Garden* by Frances Hodgson Burnett, a cheerful robin led the heroine to the door of a wall surrounding an abandoned garden that was filled with mystery and wonder. Perhaps no other animals have been as symbolically linked with the seasonal rhythms of traditional rural life as the wren and robin. Today their use, especially on greeting cards, is increasingly nostalgic.

SPARROW

> *The brawling of a sparrow in the eaves,*
> *The brilliant moon and all the milky sky,*
> *And all the famous harmony of leaves,*
> *Had blotted out man's image and his cry.*

> —W. B. YEATS, "The Sorrow of Love"

The diminutive size of sparrows has often made them objects of affection, but their raucous, noisy behavior has hurt their reputation. The Greek poetess Sappho wrote of sparrows drawing the chariot of Aphrodite, the goddess of love. The sparrow was, however, often a symbol of profane love, which was sometimes contrasted with the chaste passion of the dove. The Roman poet Catullus wrote a tender elegy beginning "Mourn ye Graces . . . " (*Lugete, o Veneres . . .*) to the pet sparrow of his mistress. The author described how the sparrow would "chirp to his mistress alone," and celebrated the great love they had for one another, clearly identifying with the bird.

Christ told his apostles, "Can you not buy two sparrows for a penny? And yet not one falls to the ground without your father knowing" (Matthew 10:29). Bede used similar imagery during the early eighth century in his *Ecclesiastical History of the English People*, in which a noble said to King Edwin, "The present life of man, o king, seems to me, in comparison of that time which is unknown to us, like the flight of a sparrow through the room wherein you sit at supper in winter . . . " He went on to compare the sparrow flying from the warm room into the wintry storm with the passage of a soul into eternity, and then to say that Christianity offered promise of certainty in that precarious journey. The way sparrows nest in almost any enclosed space including the corners of barns or porches, has also made them symbols of domesticity. Folk belief, especially in Britain, often holds that deceased ancestors may come back as sparrows.

But the sparrows of folklore can also be malignant. One European legend says that when Christ was hiding from his pursuers sparrows betrayed him by their chirping. A similar legend says that when Christ was on the Cross the swallows tried to prevent his enemies from inflicting further torments by saying, "He is dead," but the sparrows replied, "He is alive."

Their integration into the routines of everyday life made the sparrow a politically charged theme in the middle of the nineteenth century. Sparrows from England were imported and became naturalized in major cities of the Eastern United States. An intense debate known as "The Great English Sparrow War" raged for several decades as to whether these birds were harmful to American landscapes, which closely paralleled the current disputes about whether the United States should welcome immigrants. Their detractors used the same sort of rhetoric to describe the sparrows that had been used to attack foreigners, calling the birds loud, unclean, and promiscuous. Others found the little creatures a charming addition to urban landscapes. English sparrows have at last been fully accepted by Americans, though more recent additions to North America such as starlings—first released in 1890 in New York's Central Park—are still resented.

Japanese pen and ink drawing of a bat.

16
WEIRD AND WONDERFUL

WHY DO PEOPLE LOVE HORROR MOVIES? CREATURES WITH HUGE fangs and gleaming eyes may frighten us physically, but they reassure us intellectually, since we know exactly what these features mean. The beings that seem most unsettling are those that elude our habitual classifications. Their appearance may alternately suggest deities or demons, enchanted humans or dumb beasts, reptiles or mammals, and friends or foes. In *Purity and Danger*, anthropologist Mary Douglas looks at the example of the pangolin—an anteater which has scales like a fish, looks like a lizard, climbs trees like a monkey, and has warm blood. It is the center of a cult among the Lele, a tribe in Central Africa, whose members eat it only on ritualized occasions, in hope that the magic of the animal will pass to them. The animals in this section may be more familiar, but they remain just as uncanny and mysterious.

BAT

But when he brushes up against the screen
We are afraid of what our eyes have seen
For something is amiss, or out of place
When mice with wings can wear a human face.

—THEODORE ROETHKE, "The Bat"

Bats have always provided a problem for those who like to divide things into neat, unequivocal categories. Not only are they nocturnal but they also seem, in other ways, to reverse what we often assume as the natural order. They sleep hanging upside down by their feet. They live in shelters such as caves or hollow trees, but they also take advantage of human structures. Like most small animals that are often drawn to human habitations, bats have often been identified in folk belief with the souls of the dead. In cultures that venerate ancestral spirits, that has often made them sacred or beloved. When spirits are expected to pass on rather than return, bats appear as demons or, at best, souls unable to find peace.

According to one popular fable, popularly attributed to Aesop, the birds and beasts were once preparing for war. The birds said to the bat, "Come with us," but he replied, "I am a beast." The beasts said to the bat, "Come with us," but he replied, "I am a bird." At the last moment, a peace was made, but, ever since, all creatures have shunned the bat. The earliest version of this story, by the Roman Phaedrus, contained no explicit moral, and perhaps he intended to suggest that bats prefer human civilization to nature. The learned folklorist Joseph Jacobs, however, appended the lesson, "He that is neither one thing nor the other has no friends."

Today taxonomists place bats in a separate order of mammals, but both lay people and scientists have puzzled for centuries over whether bats were avians, flying mice, monkeys, or something altogether different. A revulsion against them, however, is far from universal, and their quizzical faces have often inspired affection. Glass windows had been first invented by the Romans about 400 BCE, but they remained a only a curiosity until well into the Christian era, and so most people had little choice but to share their homes with bats. According to Ovid, the daughters of Minyas had refused to join the revels in honor of Bacchus but stayed at home weaving and telling stories. In punishment, they were turned into bats, in which form they continue to avoid the woods and flock to houses. In a similar spirit, the Medieval bestiaries praised bats for the way they would hang together "like a cluster of grapes," adding "And this they do from a sort of duty of affection, a kind which is difficult to find in man." In Medieval times it was not uncommon for the entire household from the Lord and Lady to the serfs to sleep in the great hall of the manor, and little privacy was available.

In such close quarters, they must have felt rather like bats in a cave.

In Africa, Swahili-speaking people have believed that after death the spirit of the departed hovers near his body as a bat. People in Uganda and Zimbabwe have traditionally believed that bats taking wing in the evening are departed spirits coming to visit the living. The people of Ghana, however, consider the flying fox, a large bat, a demon in league with witches and sorcerers.

Perhaps the most unequivocally favorable view of bats has been found in China, where the word for "bat" also means "joy." In ancient times, the Chinese had already noticed the service that bats provided by eating insects, thus impeding the spread of malaria. Bats seemed to exemplify such Confucian virtues as filial piety, since they would live together in a single cave for generations. The Chinese believed that bats live for centuries, and Shou-Hsing, the god of long life, is still depicted with two bats.

Bats did not really come to be thought of as "spooky" in Europe until the end of the Middle Ages. Within a century or so afterwards, however, Europeans began to regard bats as familiars of witches and as a frequent disguise for the Devil. Borrowing from the iconography of Chinese art, though often giving it another meaning, Western artists began to depict dragons and demons with the wings of a bat.

The association of bats with vampires—that is, the living dead—only goes back to the latter half of the eighteenth century, when the zoologist Buffon examined newly discovered bats from South America. Because the variety sucked small quantities of blood from cattle, though rarely from human beings, he called them "vampires." At about the same time, a fashion for gothic horror began to spread through Europe, and popular writers discovered it was piquant to identify vampires with bats.

It was only in the twentieth century that the new medium of movies really established the popular association between the two. In a wave of vampire movies starting in the 1920s, actors such as Bella Lugosi gave Dracula and other vampires the appearance of bats, in many instances the ability to change into bats as well. They would, for example, sport a long black cape (like the wings of a bat), large ears, and claws. While some vampires were purely evil, others were grandly tragic, and quite a few were not so different from ordinary human beings. In the 1950s, the popular comic-book character Batman assumed many of the paraphernalia of vampire bats, but he used these to fight crime rather than to capture souls.

But the mystery of bats has not been diminished by either fantasy or science. In the late twentieth century, the philosopher Thomas Nagel probed the nature of consciousness in a famous essay entitled "What is it like to be a bat?" He tried to imagine what it might be like to navigate by sonar and decided that the human mind was unequal to the task. His conclusion was that we must recognize realities that we can neither state clearly nor comprehend.

FROG AND TOAD

> *The old pond—*
> *A frog jumps in,*
> *Sound of water.*

—BASHO (Translated by Robert Hass)

Frogs and toads are usually found around ponds or in moist areas that, in the context of myth, suggest the chaos out of which living things were created. In mythologies throughout the world, frogs are associated with the primeval waters out of which life arose. Among the Huron Indians and other Native Americans, the frog is often a bringer of rain. The aborigines of Queensland, Australia have a legend that a frog once swallowed all of the waters on the Earth. There was a great drought, and the animals decided they could only save themselves by making the frog laugh. He remained unmoved by their comic routines until the eel danced before him, awkwardly twisting and turning, at which the frog began to chortle hysterically, releasing the lakes and rivers.

When people observed their copulation, which often lasted several days, frogs seemed to embody fertility. The female frog will often lay tens of thousands of eggs every year. The Egyptian hieroglyphic sign for "one hundred thousand" was a tadpole. On hatching, the transformation of a tadpole to a frog, the major model for shape shifters in myth and legend, begins. Until the twentieth century, Europeans widely believed that frogs were generated spontaneously out of earth and water, and that frogs could survive for centuries in stone. Evolutionary theory partially confirms the intuition of early mythologists about the primal origin of frogs, since fossils of frogs have been found going back at least 37 million years.

The economy of ancient Egypt was centered about the Nile River, which teemed with frogs. The frog was particularly identified with Hekat, a deity of fertility and childbirth. When the waters of the Nile receded, innumerable frogs would be heard croaking in the mud, an event that may have influenced many myths. In one Egyptian creation myth, Heket and her ram-headed husband Khnum made both gods and human beings. According to another Egyptian creation myth, the original eight creatures were frogs and snakes that carried the cosmic egg.

The Hebrews, who reacted vehemently against Egyptian traditions, considered the frog unclean. The Bible tells us that when the Pharaoh refused to let the people of Israel leave Egypt, Yahweh sent Moses to him with this threat, which he later carried out:

> . . . know that I will plague the whole of your country with frogs. The river will swarm with them; they will make their way into your palace, into your bedroom, onto your bed, into the houses of your courtiers and of your subjects, into your ovens, into your kneading bowls. The frogs will even climb all over you, over your courtiers, and over all your subjects (Exodus, 7:27-29).

Though most of Hebrew tradition regards these animals as repulsive, some rabbinical interpreters during the Diaspora have seen the frogs in Egypt as heroic defenders of the faith. Not only did frogs fight on the side of the Hebrews, but they even hopped into fires, where they were martyred by being burned alive.

Since frogs are generally seen after a deluge, it may be that the Biblical plague of frogs once referred to a storm. The scriptures of the Zoroastrians mention frogs as among the first creatures to emerge from dark waters as part of a plague created by the demonic Ahriman. The Book of Revelations reports that demons in the form of frogs will spring from the mouths of the Dragon, Beast, and False Prophet. In Early Modern Europe, there were many reports of frogs or toads emerging from the mouth of a woman, an event that might sometimes be understood as evidence of witchcraft and, at other times, simply as a noteworthy phenomenon. In Charles Perrault's tale "The Fairies" (1697), a fairy's blessing on a kind, industrious girl, made diamonds and roses fall from her mouth with every sentence; the same fairy's curse on the young lady's lazy sister made vipers and toads drop from her lips instead. The good girl then married a prince, while the bad girl was driven from home and died alone in the forest.

The croaking of frogs around a pond has often suggested an assembly, which is a frequent theme in fables attributed to Aesop. In one of the most famous, the frogs were thriving in their swamp but longed for a conventional government. They asked Zeus to give them a king, and he laughingly threw down a log. After recovering from their awe at the splash, the frogs began to dance on the log. After a while, however, they decided the log was not a proper king and asked Zeus for another. The god, in irritation, sent down a stork, and the new monarch immediately began to gobble up the frogs.

The Classical Greek poem "The Battle of the Frogs and Mice" is a burlesque of the martial epic. It began as a mouse, fleeing a cat, took refuge in a pool. A frog offered to carry the mouse to safety but drowned him, setting off a war between frogs and mice, filled with reckless courage and bombastic rhetoric. Finally, the mice appeared on the point of victory, but Zeus, looking down with pity on the frogs, sent crabs to drive the mice back in confusion. For a long time, the poem was attributed to Homer himself, but most scholars now date it from the third century BCE. The story may have been a satire on the Peloponnesian War, in which the frogs represented Athens, the great sea power, and the mice represented Sparta, the dominant military force on land. The crabs, then, would have represented the city of Thebes, which finally broke the power of the victorious Spartans. No doubt somebody who had grown tired of the grandiloquent speeches that always accompany bloody rampages wrote "The War Between Frogs and Mice," and it is not very hard to guess why he or she may have wished to remain anonymous.

In "The Frogs" by the Athenian dramatist Aristophanes, frogs provided a chorus by croaking in the River Styx, as the god Dionysius visited the underworld—"bre-ke-ke-kex, koax, koax!" Just as the chorus in Greek tragedy tended to provide the perspective of ordinary people as counterpoint to the grand dramas being enacted, the chorus of frogs was a reminder of limitations that not even the gods could escape.

In Greco-Roman culture, frogs often had a reputation for coarseness, due to their wide mouths and muscular bodies. In his *Metamorphoses*, Ovid told how Latona, exiled from Heaven by Juno, wandered with her children, the god Apollo and the goddess Diana, about the earth. She knelt down to drink water at a stream, but some country bumpkins tried to forbid her. When she pleaded with them, the rough fellows responded with threats and insults, and they muddied the water with their feet. At last, she cursed them, saying, "Live forever in the foul puddle!" and the ruffians turned into frogs. The story anticipated the frequent depiction of toads and frogs in Hell—a favorite motif of many artists of the late Middle Ages and Renaissance.

Frogs and toads were used in innumerable medicines and magical formulas. The Renaissance zoologist Topsell reported a superstition that, if a man wished to know the secrets of a woman, he must first cut out the tongue of a living frog. After releasing the frog, the man had to write certain charms upon the tongue and lay it on the woman's heart. Then he might begin to ask questions and would get nothing but the truth. This, however, was a bit much even for the normally credulous clergyman, who remarked, "Now if this magical foolery were true, we had more need of frogs than of Justices of the Peace."

The popular distinction between frogs and toads is not fully recognized by professional biologists. Both groups of amphibians are members of the order *anura*, and they are almost interchangeable in myth and legend. The word "toad" is generally, though not always, used for creatures of the family *bufonidae*, which have short legs, rough skins, and spend much of their time on land, and they are especially associated with cultivated places such as gardens.

Toads were frequently mentioned in witch trials as familiars, as well as ingredients in witches' brews. Hieronymus Bosch showed a damned woman copulating with a toad in his painting "The Seven Deadly Sins."

Many painters of the early modern period showed demons forcing the damned to eat toads in Hell. Others depicted anthropomorphic toads or frogs cooking and devouring human beings.

The number of superstitions connected with frogs and toads was virtually endless. At least since Pliny and Aelian, toads have been widely considered poisonous. In Shakespeare's *Richard III*, the wicked king was called a "poisonous hunch-backed toad." It has, however, also been popularly believed that toads had a precious stone inside their heads. This stone has been avidly sought by alchemists for its magical properties, especially for use in detecting or neutralizing poison. Up through most of the nineteenth century and even afterwards, many books of natural history reported that frogs could survive for many centuries encased in stone.

In China, toads were one of the five venomous animals, together with the scorpion, centipede, spider, and snake. A three-legged toad was often depicted on the moon, with one leg representing each of three lunar phases. According to legend, the hermit Liu Hai decontaminated a pool by luring out the toad with a string of gold coins. He killed the toad, thus punishing the sin of avarice. Nevertheless, Liu Hai has often been painted with Ch'an Chu sitting affectionately at his side as a sort of pet, and the toad with a coin in its mouth has become a symbol of good fortune.

Sometimes people have envied the ability of frogs and toads to find contentment in a humble pond. The eighteenth century CE Japanese philosopher Ando Shoeki, a relentless critic of human arrogance, wrote that the toad once prayed to walk upright like a person. This was granted to him, but then he found that, with his eyes focused only on heaven, he could no longer see where he was going. When he regretted his request, heaven returned him to his original state. He had, the toad explained, been like the sages such as Shakyamuni or Lao Tzu, for "Looking only to the heights, they failed to see the eight organs of their own senses . . ."

In the modern period, people often found the perceived homeliness of frogs and toads endearing. An old story of a witch with a frog as her demonic companion evolved in the oral tradition to become "The Frog King" (or "The Frog Prince"), the first in the famous collection of German fairy

Nineteenth-century illustration to a fable by La Fontaine, "The Frog Who Wanted to Be as Large as a Bull."

(*Illustration by J. J. Grandville, from* Fables de la Fontaine, *11839*)

tales by the Grimm Brothers. It told of a talking frog that was disenchanted, revealed to be a prince and married to a lovely young girl. "Frog Went A-Courting" became one of the most popular of British and American folksongs. It told of a wedding feast of a frog and his mouse-bride, together with all the animals invited as guests. In Kenneth Grahame's *The Wind in the Willows* (1907), Toad is a rich, arrogant character, who, after being chastised in many rollicking misadventures, reforms to become a proper Edwardian country squire. In *The Call of the Toad*, a novel written by the German Günter Grass as a sort of epitaph for the twentieth century, the primeval voice of a toad served much the same role as the chorus of frogs did for Aristophanes—an admonition against hubris.

THE HYENA

> *I will neither yield to the song of the siren nor the voice of the hyena . . .*
> —GEORGE CHAPMAN, "Eastward Ho" (Act V, Scene I)

If Hyenas are "laughing," what is the joke? Maybe it is our efforts to put everything into neat categories. As intermediate creatures somewhere between felines and canines, hyenas have always filled people with consternation. Legends also make their gender indeterminate, and these have some basis in observation. Reversing the pattern found in most mammals, the female hyenas tend to be a bit larger than the males. More remarkable is that the females have a bulge of skin that resembles a male sex organ. The Greco-Roman author Aelian claimed that the hyena changed its sex every year. According to Aelian, the hyena could also put a creature to sleep with a mere touch of its left paw. The shadow of a hyena cast against the full moon reduced dogs to silence, so they could be carried off without resistance. The hyena imitated the human voice in order to lure dogs and even men to their doom. In seventeenth-century England, Topsell reported a rumor that hyenas could be impregnated by the wind.

From Africa to Europe, the hyena has usually been considered treacherous, stupid, and cowardly. One possible reason is that hyenas occupy roughly the same habitats in Africa as lions, and they often live by scavenging. Lions are symbols of kingship, making hyenas seem like venial courtiers. According to Medieval European bestiaries, the hyena (sometimes called the "yena") lived in tombs and devoured dead bodies. Bestiaries also reported that the hyena had a stone in its head, which, when taken out and placed under the tongue of a person, could enable a man or woman to see the future.

One rare legend makes the hyena not only gentle but also motherly. Saint Macarius of Alexandria lived as a hermit in the desert, where his skill in healing was known even to the animals. Once, a mother hyena came to him bearing her baby in her mouth. Saint Macarius picked up the infant, looked it over, and realized that it was stricken with blindness. He made the sign of a cross with one hand over the baby hyena's eyes, at which it immediately went to its mother's breast and began to suck. Later, the mother hyena brought Saint Macarius the skin of a sheep, freshly killed, in gratitude. Saint Macarius, troubled by the mother having killed the sheep, refused to accept the gift. The mother bent her head and entreated. Finally, the holy man agreed, but only if the mother would promise henceforth not to hurt the poor by taking their sheep but to take only meat that was already dead.

When the mother assented, he took the sheepskin and slept upon it until his death.

In the twentieth century, other maligned animals such as wolves have been redeemed, but the reputation of the hyena has changed little. In the 1994 Disney film *The Lion King*, the evil lion that wishes to claim the throne aligns himself with the hyenas. The makers of the film wished to be ecologically sensitive and to avoid condemning any animal, so they usually made the hyena leaders more goofy than malign. Nevertheless, the hyenas advancing against the lions in step recalled the Nazis marching in Leni Riefenstahl's Nazi propaganda film *The Triumph of the Will*. Several naturalists, perhaps beginning with Saint Macarius, have pleaded with the public without success to redeem the reputation of the hyena.

MANTIS

> *A grey locust, heedless of danger, walks towards the mantis. The latter gives a convulsive shudder, and suddenly, in the most surprising way, strikes an attitude that fills the locust with terror, and is quite enough to startle anyone.*
>
> —HENRI FARBE, *Farbe's Book of Insects*

The word "mantis" is Greek for "seer," and the praying mantis indeed looks the part. With its green or brown color, diaphanous wings, and stick-like appendages, the mantis blends very easily into foliage, but its long front legs are constantly moving. When they are raised, the mantis appears to be making a gesture of supplication; when they are lowered, it suggests hands folded in prayer. The wings of a mantis resemble the flowing robes of a priest, but the most noticeable feature of the insect is its enormous eyes, to which ancient legends accord a power to curse.

Among the Ngarinyin, an aboriginal people of Northern Australia, the mantis is the symbol of the mother goddess Jillinya, and it is often painted on rocks and in caves. Parents warn their children not to harm mantises, for Jillinya might retaliate by bringing storms. Among the Kalahari Bushmen of Southern Africa, the mantis is also regarded as a creator deity,

A praying mantis reborn into a better life.

(Illustration by J. J. Grandville, from Scènes de la vie privée et publique des animaux, *1842)*

which controls the wind and rain. People of East Asia have been impressed by the aggressiveness of the praying mantis, which is willing to attack creatures several times its size. A traditional saying in Japan goes, " . . . like a mantis raising its arms to stop the wheel of a passing cart."

The mantis of folklore seems to veer between extremes of good and evil. There is a widespread legend that the praying mantis can divine the goal of a traveler at a glance, as well as any possible dangers along the way. In the mid-seventeenth century, Thomas Muffet (writing in collaboration with Edward Topsell) wrote of the mantis, "So divine is this creature esteemed, that if a child ask the way to such a place, she will stretch out one of her feet, and show him the right way, and seldom or never miss." He added that mantises, ". . . do not sport themselves as others do, nor leap, nor play, but, walking softly, she retains her modesty, and shows forth a kind of mature gravity."

In the nineteenth century, however, when females of the species were observed to eat their mates after copulation, people were disillusioned and began to demonize the mantis. The act could, however, also be interpreted romantically, by saying that the male mantis makes the ultimate sacrifice, in order to truly become one with his beloved.

Farbe remarks in this *Book of Insects* that the mantis "seemed like a priestess or nun," but immediately added:

> There was never a greater mistake! Those pious airs are a fraud; those arms raised in prayer are really the most horrible weapons, which slay whatever passes within reach. The mantis is fierce as a tigress, cruel as an ogress. She feeds only on living creatures.

He goes on to describe how the mantis can terrify even larger insects such as locusts. But it is hard to mistake a note of admiration for the mantis in Farbe's description, not only for its martial prowess but also for courage and integrity.

Many philosophers, theologians, anthropologists and others have seen a kinship between the sacred and the profane. Deities from Ishtar to Yahweh have often been objects of love and terror, often of both at once, and no other animal exemplifies that union more than the mantis. In the words of Charbonneau-Lassay, the mantis is "the image of Christ, the guide of souls," who may bring people to either Heaven or Hell.

Crocodile

17
BEHEMOTHS AND LEVIATHANS

VERY FEW IF ANY BIBLICAL PASSAGES ECHO THROUGH FOLKLORE AS PERSIST-ently as the speeches in which Yahweh reminds Job of the grandeur of Creation, especially Behemoth (Job 40: 15-24), a giant creature that walks the earth, and Leviathan, a sea monster. Both are described through long sequences of majestic comparisons, which communicate enormous grandeur but little precision. Since we cannot tell what existing animals, if any, Behemoth and Leviathan were intended to represent, they have been associated with a range of creatures on both land and sea. "Behemoth" has been taken to be an elephant, hippopotamus, rhinoceros, or depicted as a composite of the three; "Leviathan" is usually thought of as a crocodile or whale. The feature that stands out in all legends, pictures, and descriptions of them is their enormous size, yet their splendor transcends measurement.

CROCODILE

> *How doth the little crocodile*
> *Improve his shining tail,*
> *And pour the waters of the Nile*
> *On every golden scale!*
> *How cheerfully he seems to grin,*
> *How neatly spreads his claws,*
> *And welcomes little fishes in*
> *With gently smiling jaws!*

> —LEWIS CARROLL, *Alice's Adventures in Wonderland*

Crocodilians—which include crocodiles and alligators—are, together with komodo dragons, the only large, partially terrestrial animals that do not hesitate at all to attack and devour human beings. What makes crocodiles especially frightening is the suddenness with which they strike. Most of the time, they appear utterly lethargic, yet they can rouse themselves almost instantly and attack, for short periods, and with remarkable speed. Sometimes a lunge will thrust a crocodile partially out of the water until, for a second or so, it is almost standing upright.

Crocodiles still inspire a sort of primeval terror, yet they do not seem entirely alien to us. The expressions in the eyes of most reptiles seem almost impossible for us to read, but those of crocodiles sometimes appear to share a glimmer of human awareness. Female crocodiles care briefly for their young and, according to some observers, crocodiles may even engage in cooperative hunts. The upturned mouth of a crocodile can appear to be a perpetual smile, but the large teeth that always protrude on the sides give it a sinister aspect.

Crocodiles are closely identified with wetlands and, in consequence, with irrigation and fertility. According to legend, Menes, the first king of Egypt, was hunting when he fell into a swamp. His dogs failed to help him, but a friendly crocodile ferried the monarch to safety on its back. On the place where he arrived in safety, Menes founded the city of Crocopolis, where the crocodile-god Sebek was worshipped. Much the same story was later told of Saint Pachome, who founded a monastic order in Egypt during the third century. He was so beloved by animals that crocodiles would ferry him across the Nile River to whatever destination he might request.

The Greek historian Herodotus reported that Egyptians in some districts killed and ate crocodiles, but those in others considered the animals sacred. In Crocopolis, priests would place one (presumably) tame crocodile in a temple, where golden ornaments would be placed in its ears and bracelets on its legs. Pilgrims would bring the holy crocodile special offerings to eat, and, after death, it would be embalmed and placed in a coffin. Herodotus, who visited the labyrinthine temple containing the remains of crocodiles and kings at Crocopolis, wrote, "Though the pyramids were greater than words can tell . . . this maze surpasses even the pyramids."

Illustration entitled "The Lying Demon" from a religious tract, Philadelphia, ca. 1900. The deceiver is accompanied by a crocodile, symbol of hypocrisy.

Other mythologies throughout the world reflect admiration for the crocodile and its power. The dragon of Chinese mythology—which appeared to the Emperor Fu Hsi out of the Yellow River— resembles a crocodile with its teeth and short legs, though stylized almost beyond recognition. A Muslim legend from Malaysia holds that Fatima, daughter of Mohammed, created the first crocodile. In some parts of Java, mothers would traditionally wrap the placentas of their children in leaves and place them in a river as an offering to ancestral spirits that have become crocodiles.

But terror of, and scorn for, the crocodile go at least equally far back in history. In paintings to illustrate *The Egyptian Book of the Dead*, the goddess Ammut would be shown waiting hungrily to devour those who were found wanting, as a soul was weighed in balance. In this capacity, she had the head of a crocodile, as well as the forepart of a lion and the hind legs of a hippopotamus. Several monsters of legend, to which human sacrifices were made, may have been originally crocodiles. In Greek mythology, for example, the Ethiopian maiden Andromeda was chained to a rock to be eaten by such a creature, before the hero Perseus saved her. Human sacrifices to crocodiles—made by chaining the victim beside a lake or river—have been widely practiced from Africa to Korea.

The crocodile has been closely associated with magic from time immemorial. In one Egyptian text from the early second millennium BCE, a sorcerer makes a wax crocodile and throws it into the Nile River. It immediately grows large and devours the lover of his wife. Sorcery is always closely linked with deception, and in Western Europe, the crocodile has been a symbol of hypocrisy. Medieval Bestiaries reported that crocodiles weep as they eat human beings. Topsell wrote in 1658 of the crocodile that "to get a man within his danger, he will sob, sigh, and weep, as though he were in extremity, but suddenly he destroyeth him." The early naturalist also noted that, according to other observers, the crocodile wept after eating a man, much as Judas had cried after betraying Christ.

Several peoples, including Arabs and some African tribes, have offered accused criminals to crocodiles as a test, and those people who were eaten or bitten were presumed to be guilty. In the Middle Ages, the entrance to Hell was sometimes depicted as a huge jaw filled with teeth, which might be that of a lion, snake, or crocodile. The idea that crocodiles only eat the guilty has persisted into the latter twentieth century among the Turkana people who live around Lake Rudolph in Kenya. When Alistair Graham saw them wading casually into waters filled with crocodile, he was told by a tribesman, "My conscience is clear; therefore, I am in no danger."

In the British children's classic *Peter Pan*, James M. Barrie created the villain Captain Hook (named for an iron claw that has replaced one hand), a hypocritical murderer like the crocodile of legend. His hand was once bitten off by a crocodile, which liked the morsel so much that he followed Hook ever since, though the beast does not threaten anybody else. The crocodile had also swallowed a clock, and the Captain was terrified whenever he heard it tick. The beast represents the ravages of time, which Peter himself has escaped by refusing to grow up and journeying to Neverland. When Hook was finally kicked overboard to the crocodile, the clock stopped and Hook went contentedly to his death, a bit like the victim of a human sacrifice who believes that to be eaten is an exalted destiny.

ELEPHANT

Nature's great masterpiece, an elephant; the only harmless great thing.
—JOHN DONNE

The elephant is set apart from other creatures by its immense size, its enormous tusks, and, above all, its prehensile trunk. Rough as the skin of an elephant may appear, the trunk has such fine coordination that it can be used to pick flowers or lift small coins. But this strange, paradoxical nature has made people identify intensely with the elephant, since the animal seems to share human alienation from the natural world. Cicero wrote in the first century BCE, " . . . although there is no animal more sagacious than the elephant, there is also none more monstrous in appearance." No other animal has been so intensely and consistently anthropomorphized. The eyes of an elephant are disproportionately small and on opposing sides of the head, but the folds about the eyes give them enormous expressiveness.

Pliny the Elder spoke for many when he said the elephant was the animal "closest to man as regards intelligence," adding that "the elephant has qualities rarely apparent even in man, namely honesty, good sense, justice, and also respect for the stars, sun, and moon." One traditional description of humanity is *homo religiosus*, but elephants, according to tradition, even share the human religious impulse. Pliny also wrote that elephants would come from the mountains down to Mauritania to bathe in the river Alimo and pay homage to the moon. This was frequently repeated in Christian Europe, where the religiosity of the elephants was regularly praised and their paganism was ignored. In the latter eighteenth century, Marcel LeRoy, forester to King Louis XV of France, would write, " . . . many authors say this animal is lacking in nothing but the worship of God, while others accord it that virtue as well." Even today, the debate as to whether elephants are religious has not been entirely resolved. For centuries, elephants have been said to bury their dead, and researchers in the latter twentieth century confirm that they at least cover their dead up with vegetation.

Most of our lore regarding the elephant comes from India, where the elephant may have been domesticated as early as the middle of the third millennium BCE. After the sun had been hatched from a comic egg, the god Brahma took the two shells in his hands and began to chant. Out of the shells emerged the elephant Airavata, which later became the mount of the god Siva, followed by fifteen other cloud elephants. They and their progeny could fly about and change their shape at will. One day, however, the young elephants on Earth became too boisterous and disturbed the sage Palakapya, who cursed them and confined them to the ground.

Up through at least the nineteenth century, elephants have continued to be accorded superhuman abilities, including total recall and life spans of centuries. But even as these ideas have been debunked, remarkable new qualities of elephants have been discovered. People had long puzzled over the social cohesion of elephants, and in the 1980s researchers discovered that elephants communicate with ultrasound—that is, by means of frequencies inaccessible to the human ear.

Among the most beloved deities of the Hindu pantheon is Ganeša, the mischievous Hindu god of wisdom with a human body and an elephant head. He is usually depicted riding on a rat, has a potbelly and one broken tusk. There are many stories that explain his origin and odd appearance. In one of them, when the god Siva was on a journey, his consort Parvati became lonely and desired a son. She covered her body with scented lotion, rubbed off the dirt, formed it into a young man, and directed him to guard her chamber. After a while, Siva returned and demanded admittance, but the young man refused to let him pass. A fight ensued, and Siva beheaded his adversary. When she saw what has happened, Parvati became so furious that she threatened to destroy the entire world if Siva did not restore her son to life. To do this, Siva needed another head, so he sent his servants in search of one. They came upon an enormous elephant, decapitated the animal, and return to their master, who placed the head of the elephant of the body of the young man.

One of the many incarnations of Buddha was an albino elephant, and such animals are traditionally held in great honor in Southeast Asia. The conception of Gautama, who was to become Buddha, resembles that of Christ, but the mediator of that virgin birth was not a dove—symbol of the Holy Spirit in Christianity—but an elephant. Queen Sirimahamaya dreamed that she had been transported to a palace on a mountain peak. An elephant, bearing a lotus, approached her and bowed. She heard the call of a bird and awoke, pregnant with the redeemer.

The elephant entered European awareness at the battle of Gaugamela when Alexander the Great invaded India and faced King Porus, whose army included 200 mounted elephants. We only know what happened from Western accounts written hundreds of years later, but they state that Alexander was finally victorious. They add, however, that the power of the elephants so awed his troops that his generals refused to venture any further eastwards. At the end of the third century BCE, Pyrrhus, the king of Epirus in Greece, used elephants to defeat the Romans in several battles before the superior numbers of his adversary finally overwhelmed him. In 219 BCE the Carthaginian general Hannibal crossed the Alps with an army that included elephants and inflicted many defeats on the Romans, though his troops also finally succumbed to the superior numbers and discipline of the Romans. It is interesting that the Romans, despite their experience of the damage elephants could inflict in battle, only very rarely included these animals in their legions. Perhaps the reason is that the Romans loved order, and elephants, even when trained, were too unpredictable for them in battle.

The reason may also be that the Romans were too fond of elephants to use them in such a manner. They were slain in Roman circuses, but the spectacles were not very popular. Pliny also recorded that, on seeing one elephant killed with a javelin in the area at a festival organized by Pompey, the others tried to break through iron railings. "But when Pompey's elephants had given up hope of escape, they played on the sympathy of the crowd, entreating them with indescribable gestures. They moaned as if wailing, and caused the spectators such distress that, forgetting Pompey and his lavish display devised so to honor them, they rose in a body, in tears, and heaped dire curses on Pompey . . . " The resulting loss of popular support,

Pliny believed, was partly responsible for Pompey's defeat by Caesar not long afterwards.

Haroun al Rashid gave an elephant as a gift to Charlemagne and his court. A few European princes of the late Middle Ages also imported elephants as exotic trophies, to display their splendor and power. In general, Europeans of the Middle Ages knew elephants only through ancient books and confused reports by mariners, but the animals were far from forgotten. As their physical presence vanished, their symbolic importance increased, and no Renaissance painting of the Noah and the Flood or of the Garden of Eden was complete without a pair of elephants.

According to a Medieval bestiary, elephants lived for hundreds of years. They also related that, when an elephant couple wanted to have a child, they would go eastwards towards Paradise until they came to the Mandragora, the Tree of Knowledge. First, the elephant wife would eat from the tree. Then she would give some of the fruit to her husband, at which the two would copulate and immediately conceive. The elephants were like the first Adam and Eve, except that the fruit from the Tree of Knowledge was not forbidden to them, and they might freely enter or leave the Garden of Eden. The author also says of elephants, "They never quarrel with their wives, for adultery is unknown to them. There is a mild gentleness about them, for, if they happen to come across a forwarded man in the deserts, they offer to lead him back to familiar paths."

In most of the Arab world, the elephant was only slightly less exotic than in the West, but it was held in much the same high regard. The seventh voyage of Sinbad the sailor, from the Medieval *Arabian Nights Entertainments*, contains an episode that anticipates modern ecological and humane concerns. Sinbad had been captured by pirates and sold into slavery. At the direction of his new master, he hid in a tree and shot arrows at a herd of elephants, in order to obtain their tusks for ivory. This continued for a few days, but then the elephants surrounded him and uprooted the tree. Sinbad expected the elephants to kill him, but instead they took him to their graveyard, so he might peacefully obtain their ivory.

Meanwhile, in Africa, elephants were very much a physical reality, and the practical problems of living alongside elephants restricted their appropriation in fantasy or symbolism. They were a copious source of meat but also a formidable challenge to hunters. An Ashanti proverb goes, "If you follow an elephant, you don't have to knock the dew from the grass." Strength and power rarely go together with cunning in folklore, and African tales often present the elephant as mighty but naïve. According to a Central African tale from the Mbochi people, the animals selected the elephant as their king. As he was going to his coronation, the hare lay in his path and pretended to be terribly ill. The elephant did not wish the hare to miss the great event, and so he lifted the little fellow upon his back. When they reached the council of animals, the hare protested that the elephant had carried him on his back—as the rider, and so he must be superior to his beast of burden. The animals crowned not the elephant but the hare as king.

Africans traditionally hunted elephants primarily for meat, so both the danger and the bounty obtained from a single kill kept slaughter within limits. With colonization and the advent of modern weapons, the demand for ivory has placed the once vast populations of elephants in danger. Desperate to stop poaching, some African governments in the latter twentieth century have imposed the death penalty for killing of elephants. Elephants are accorded an ironically human status, both as objects of slaughter and protection.

HIPPOPOTAMUS

> *So he lies beneath the lotus,*
> *And hides among the reeds in the swamps.*
> *The leaves of the lotus give him shade,*
> *And the willows by the stream shelter him.*
> *Should the river overflow on him, why should he worry?*
> *A Jordan could pour down his throat without his caring.*

—JOB 40: 21-23

For most of history, people outside of central and southern Africa have known the hippopotamus mostly through vague rumors and a few Roman mosaics. Its memory, however, had been firmly implanted in the collective imagination through the words spoken by Yahweh in the Bible:

Now think of Behemoth;
> He eats green stuff like the ox.

But what strength he has in his loins,
> what power in his stomach muscles!

His tail is as stiff as a cedar,
> The sinews of his thighs are tightly knit.

His vertebrae are bronze tubing,
> His bones are hard as hammered iron.

His is the masterpiece of all God's work . . . (Job 40:10-19).

The identity of Behemoth is a perennial subject of debate, and tradition has at times made the creature an elephant, a rhinoceros, or even a crocodile. The passage went on to describe Behemoth as living beneath the lotus, something that best fits the hippo. When the book was written, the Hebrews, living in Mesopotamia, probably had little or no contact with any of those animals, and they may well have confused the various creatures already. An engraving of Behemoth by the British poet William Blake gave the monster features of hippopotamus and elephant, an idea that may be close in spirit to the Biblical inspiration.

Illustration of a hippopotamus from a nineteenth century book of natural history.

For those in more intimate contact with the hippopotamus, the animal has seemed almost as awe inspiring but less forbidding. For all their bulk and power, hippos can move with surprising agility and even a sort of grace. The Egyptian goddess Taweret was generally depicted with the head of a hippopotamus and the body of a pregnant woman. Her image was found on several amulets, and symbolized the fierceness of maternal love. The Egyptians, however, regarded the hippo with ambivalence, perhaps because it would often trample crops. The villainous god Set took the form of a red hippopotamus in his unsuccessful battle with the god Horus for control of Egypt.

In many places in Africa, people have venerated a hippopotamus goddess who resembled Tawaret. The Ronga of Southern Mozambique tell of a woman who gave her son to a hippopotamus goddess for protection. The deity raised the boy beneath the river, yet brought him up every evening to be suckled by his mother.

In the middle of the seventeenth century, the British clergy man Edward Topsell, wrote that the hippopotamus or "sea-horse" was "a most ugly and filthy beast, so called because in his voice and mane he resembleth a horse, but in his head an ox or a calf, in the residue of his body a swine . . ." Though he had little trouble accepting the existence of such creatures as the unicorn or satyr, Topsell was skeptical about that of the hippo.

In 1849 a hippo named Obaysh, the first seen in Europe since Roman times, was brought to the London Zoo, and he promptly generated a craze known as "hippomania." Thousands of people lined up to see him every Saturday, newspapers chronicled every detail of his life, and there was even a dance named the "Hippopotamus Polka." People have continued to be fascinated by the surprising agility of these enormous animals, and the film *Fantasia*, released in the 1950s by Disney, features hippos dressed as ballerinas. Whether the creators of the film were aware of it, these animals were not entirely unlike the hippopotamus goddess of ancient Egypt.

WHALE

Yahweh had arranged that a great fish should be there to swallow Jonah; and Jonah remained in the belly of the fish for three days and three nights. From the belly of the fish he prayed to Yahweh . . . .

—JONAH 2:1-2

The blue whale may be the largest animal to have ever lived, and other varieties are also huge. This can make the whale difficult to empathize with or to humanize. In myth and legend, the whale appears often, though not always, more as a force of nature than as an individual animal.

Tradition usually equates the creature known as "Leviathan" in Jewish legend with the whale, though a few people have wondered whether Leviathan could have been some other creature that is now extinct or has retired into the depths of the sea. Jewish tradition tells that on the fifth day of Creation, God made two Leviathans, male and female. They were so huge that the entire earth could rest on one of their fins. Soon God realized that, should they live to reproduce, they would soon destroy the entire universe, so he killed the female leviathan. Then, so that the race would not perish completely, God allowed the male to live until the end of time. So that Leviathan would not be lonely, God spends the final hours of each day playing with the monster. In the final days of the world, Leviathan shall do battle with Behemoth, and both shall be killed. A tent for the just shall be made from Leviathan's skin, and they will feast on the meat of both monsters.

In many Jewish legends, Leviathan was the ruler of the kingdom below the sea. In the story of "King Leviathan and the Charitable Boy," recorded by Nathan Ausubel, a young man heard the commandment, "Cast thy bread upon the waters . . ." Not understanding but wishing to do right, he went out every day and threw bread into the sea. A single fish noticed, waited every day for the boy, and ate the bread. The fish grew bigger than all the others in the sea. The smaller fish became afraid, and they complained to King Leviathan. When he leaned what has happened, King Leviathan commanded the big fish to bring the young boy. When the boy came to the sea on the following day, the big fish swallowed him, carried him to King

Whales or related animals from the Carta Marina of Olaus Magnus, published in 1539.

Leviathan, and vomited him out. The boy told King Leviathan that he had tried to follow the commandments of God. The Ruler of the Sea took the boy on as a pupil, taught him the Torah, and instructed him in every language of man or beast. When the boy finally returned home, he became a man of great scholarship and wealth.

In the Book of Jonah, Yahweh commanded Jonah to preach in the Assyrian city of Nineveh, which was full of wickedness. Jonah was afraid and took refuge in a ship, but soon there was a terrible storm. Jonah realized that the storm was sent because of him, and told the sailors to throw him overboard. When they did, they sea immediately became calm. Jonah was swallowed by a large fish, and he spent three nights in its belly. Finally, the fish vomited him up on the shore. Jonah traveled to the Assyrian capitol of Nineveh, where he converted the city and saved it from destruction. The fish was identified as a whale in Matthew 12: 40.

One traditional Jewish interpretation of the tale is that the three days spent within the fish represented the exile of the Hebrew people, who should return once again to Zion. In Christian Europe, the jaws of the whale were often used to represent the gate of Hell. The figure of Jonah, swallowed and cast forth, was understood as an anticipation of Jesus Christ, who spends three days in the grave to rise again. A late fourteenth-century English author, sometimes known as the "Pearl Poet," retold the Biblical story of Jonah in a poem entitled "Patience," in which the author compared the belly of the whale, with its filth and stench, to Hell. Jonah prayed to God, and then managed to find a clean niche where he waited for deliverance.

From antiquity through the Renaissance, many tales were told of sailors who would camp and make a fire on the back of a whale, mistaking it for an island. Medieval bestiaries reported that the whale, troubled by the heat, would dive to the bottom of the sea, drowning the mariners. They compared the victims to those foolish people who fail to recognize the work of the Evil One. The depths of the sea, filled with unknown creatures, were equated with Hell, and the whale became the Devil.

But the whale might also be benevolent. The legendary Medieval Irish abbot Saint Brenden sailed among wondrous islands with his followers for seven years in a search for the Paradise of Saints. Every year at Easter a whale named Jasconius would appear and let the monks celebrate mass upon its back. In the tall tales of the fictive Baron Munchausen, first written down by R. E. Raspe (1785) and later elaborated by several others, the Baron's ship was swallowed by a whale. Inside the monster, Munchausen found boats from around the world, some of which have been standing at anchor for many years. Under the Baron's leadership, a few brave sailors manage to prop open the mouth of the monster with a mast, enabling everyone to escape.

Native Americans of the Northwest coast have long hunted whales in shallow waters using small boats. The orca or "killer whale" is a common totem of the Haida, Tlingit, Tsimshian and related tribes, who tell stories of ancestors who put on the skins of killer whales, assumed their form, swam with them in the ocean, and sometimes married them. The whale has been central to the cultures of other tribes, including the Mahkah in Northern Canada, which revived its ancestral whaling traditions during the 1980s and 1990s.

The Taiwanese credit the whale with protecting their island by driving back the ships of foreign invaders. The Japanese maintain that, in the twelfth century CE, a governor of the Shichito islands in the southeast named Yoda Emon was saved by a whale after his boat had been destroyed in a storm. In gratitude, the governor forbade whaling in his territory.

Up until the early modern period, the whale was only occasionally glimpsed by Europeans, and then only at a distance or under tense conditions, making it an object of constant rumors and legends. By the eighteenth century, whaling had become a major industry. Whale oil was in great demand for lamps, and the highly constricting corsets worn by ladies were made of whalebone. The hunting was so intense that by the middle of the nineteenth century people already worried that the Atlantic Ocean was close to being depleted of whales.

Moby Dick by Herman Melville (1851), is an epic novel that was written to record the folkways of whalers before they vanished completely. Ahab, the captain of a whaler, had lost a leg to the great white whale named Moby Dick, and he became obsessed with his desire for revenge. He relentlessly pursued his adversary, even after the hull of his ship was filled with whale oil, until Moby Dick finally killed Ahab and destroyed his ship. The whale here, inspired at least partly by the Biblical Leviathan, punished the hubris of Ahab and all of humankind.

After whales were hunted to near extinction in the first half of the twentieth century, the ecology movement adopted the whale as a symbol. "Save the whales" was a resonant slogan that seemed to go beyond one creature and embrace the entire natural world. Whale songs, perhaps the most complex melodies produced by any creature in the wild, were recorded not only for scientific study but also for entertainment. International restrictions on whaling have now enabled them to increase their numbers to a point where many species are no longer considered endangered. Perhaps the songs may have helped?

Snowy Owls, c. 1840. The archaic divinities accompanied their more human successors, often as mascots or alternate forms. Athena, for example, was pictured with an owl.

18
DIVINITIES

T HE PREHISTORIC CAVE PAINTINGS OF FRANCE AND SPAIN ARE AMONG OUR most ancient works of art. When human beings appear in these paintings, they are usually crude stick figures that the artists must not have considered very important. The animals are painted with far more care and passion. The first clearly identifiable religious shrines in history are at Çatal Huyuk in Anatolia and date from around the middle of the seventh millennium BCE. They were dedicated to animals, especially bulls, but also vultures, foxes, and others. The Egyptians also worshipped deities incarnated in bulls, crocodiles, cats, ibises, and many other creatures. Over millennia, anthropomorphic goddesses and gods slowly replaced the animal deities. The archaic divinities accompanied their more human successors, often as mascots or alternate forms. Athena, for example, was pictured with an owl, Zeus with an eagle; Odin was accompanied by ravens and by wolves. The monkey Hanuman, who fought alongside the hero Rama in the epic *Ramayana*, is now perhaps the most popular figure in the Hindu pantheon. There is a bit of an archaic mother goddess in the "wicked witch" of Halloween with her faithful bat, spider, or black cat at her side. There has been a vast range of animals that, in particular cultures or historical epochs, have been considered divine, and this section looks only at a few that have most consistently been accorded that status.

One of the many animals that could legitimately have been included in this chapter, which we have placed elsewhere, is the lion. It may be found in Chapter 6, "Tooth and Claw."

DOVE

And the dove came to him in the evening; and, lo, in her mouth was an olive leaf pluckt off: so Noah knew that the waters were abated from off the earth.

—GENESIS 8:11

Doves seem holy and clean, but pigeons appear commonplace and dirty. Nevertheless, the two are very closely related in both biology and folklore. In ancient texts, it is often impossible to know which is meant, and perhaps the best way to think of these birds is as the sacred and profane aspects of a single creature.

A grove near the city of Dodona contained one of the most ancient and venerable oracles in Greece. In remote times, a black dove from Egypt alighted there. As it moved among the oak trees, the branches would rustle and speak to the priests with the voice of a woman. In the time of Homer, the shrine at Dodona was the most revered in all the land.

In the ancient world, doves were often associated with prophecy. In Apollonius of Rhodes' *The Voyages of the Argo*, the Greek heroes in search of the Golden Fleece found their way through the sea barred by the Clashing Rocks, which would continually open and close. They released a dove, and because it passed between the rocks, the heroes knew they could navigate unscathed. In Virgil's *The Aeneid*, doves guided Aeneas through a forest to a golden bough, which he needed in order to enter the world of the dead. Even Christianity, which often took a dim view of pagan oracles, was full of stories in which doves assisted in divination, perhaps because the doves seemed above every suspicion of evil. One apocryphal gospel told that a dove from heaven alighted on the staff of Joseph, anointing him as the husband of Mary.

Of course, whatever pleased the gods would be offered up to them in the ancient world. For the Hebrews, doves and pigeons were the only birds that might be offered for sacrifice (Leviticus 1:14), and they were the favorite sacrifice of people who could not afford sheep or oxen. The Biblical book of Genesis stated that, "God's spirit hovered over the water." This image certainly suggested a bird, and it has usually been depicted as a dove. During the Flood, Noah sent out a dove. When it returned with an olive branch, he knew that the waters had begun to subside. In Christianity, the dove represents the Holy Spirit. A dove descended on Jesus at his baptism. In pictures of the Annunciation, the dove has traditionally been portrayed descending to Mary from God the Father as she becomes pregnant with the infant Jesus. The scene recalls the amorous adventures of Zeus, for example for example when the god assumed the form of a swan to impregnate the maiden Leda. The dove, usually painted directly between Mary and God the Father, seemed to shield Mary with its purity.

The dove was sacred to many goddesses of the ancient world. Doves drew the chariot of Aphrodite, the Greek goddess of love. Though sometimes thought promiscuous, Aphrodite became a guardian of chastity when the hunter Orion attempted to break into the home of the Pleiades, the seven daughters of Atlas and Pleione. She changed the girls into doves so they might escape by flight, and Zeus later transformed them into stars.

Doves fed the legendary Assyrian queen Semiramis, daughter of the goddess Derceto, when she was abandoned as an infant in the desert. They were also closely associated with the Roman Venus, the Babylonian Ishtar and the West-Semitic Astarte. This amorous symbolism enters the Judeo-Christian tradition through the Biblical "Song of Songs," which probably referred to turtle doves:

> The season of glad songs has come,
> the cooing of the turtledove is heard.
> The fig tree is forming its first figs
> and the blossoming vines give out their fragrance,
> Come then, my love,
> my lovely one, come.
> My dove, hiding in the clefts of the rock,
> In the coverts of the cliff,
> show me your face,
> let me hear your voice . . . (2:12-14)

Jews and Christians have interpreted this song of love as an allegory of the longing of the soul for God. The image of the dove has always served to spiritualize erotic desire. It is also a symbol of conjugal fidelity. According

A Greek grave relief from the island of Paros, ca. 455–450 B.C. This girl is remembered caring for animals, perhaps at the annual festival of Artemis. She kisses one of two doves or pigeons which will go on to have a family though fate denied that privilege to her.

to the Medieval German poet Wolfram von Eschenbach in his epic *Parzifal*, written in Germany around 12000, a dove that has lost her mate would always perch on a withered branch.

The Holy Spirit is traditionally spoken of with a masculine pronoun. Nevertheless, it is very hard to think of it as male. The Trinity and the very concept of God seem unbalanced without some feminine element. Several heretical groups have identified the dove with the feminine concept of *sophia* or divine wisdom, as well as with Mary herself. The wings of a dove spread out and pointing downwards are sometimes stylized in Christian art to form an "M" for "Mary."

When Christianity was first introduced into Russia, people were forbidden to eat the flesh of doves. The dove is also important in the Grail Romances. In *Parzifal*, a dove visited the Castle of the Grail every year on Good Friday to bring the Host from heaven. The Dove was also the badge of the Knights of the Grail. European folklore made the dove the one shape that the devil could not assume. The dove was also one of the very few common animals that were never mentioned as a familiar of a witch.

In the ancient world, several cultures portrayed the soul as a dove. There is an enormously moving sculpture in the Metropolitan Museum from the grave of a Greek child who died in the mid-seventh century BCE. The young girl holds a pair of doves in her hands, and her lips touch the beak of one. The doves are to go on and have the marriage and family that were denied the maiden. The dove was the symbol of Saint Scholastica, founder of a convent and the patroness of rain. Her twin brother, Saint Benedict, visited her on her deathbed. When she died, Saint Benedict saw her soul ascend to heaven in the form of a white dove.

The dove is also holy in Islam. Christian polemicists sometimes tried to discredit Islam by claiming that Mohammed had a dove feed from his ear. This was allegedly a trick to make his followers believe that the Holy Spirit was giving him advice.

In their collection of German legends, the Grimm brothers tell how a dove saved the town of Höxter. This community had held out valiantly against the mighty army of the Holy Roman Empire during the Thirty Years' War. At last, when other attempts had failed, the imperial generals ordered their troops to bring the heavy artillery and bombard the town into submission. In the evening, a soldier was about to light the fuse of the first cannon, when a dove flew down and pecked his hand, forcing him to drop his kindling. The soldier took this as a sign from God, and he refused to fire. This delayed the bombardment long enough for Swedish troops to arrive and lift the siege.

The dove, particularly because of a drawing by Picasso, became a symbol of the peace movement during the Cold War. Is the dove a little too perfect? It suggests eroticism without lewdness and virtue without self-righteousness. It is rare indeed for any symbol to be accepted with so little ambivalence. Perhaps that is possible in this case because the pigeon functions as a sort of double to the dove, deflecting any resentment. Few people ever even think, at least consciously, of a connection between the dove on the street and the one in church, but isn't that true of many religious symbols?

EAGLE

He clasps the crag with crooked hands;
Close to the sun in azure lands,
Ringed with the azure world, he stands.
The wrinkled sea beneath him crawls;
He watches from his mountain walls,
And like a thunderbolt he falls.

—ALFRED LORD TENNYSON, "The Eagle"

The symbolism of no other animal is quite as simple and unambiguous as that of the eagle. It is associated with the sun and, largely by implication, with monarchs. Contrary to their reputation, eagles are not exceptionally high flyers among birds, but they are extremely powerful and often able to lift large prey such as sheep or monkeys. Perhaps their remoteness has also contributed to an exalted reputation, since they prefer rocky cliffs or tall trees for their nests. Though eagles may be majestic, we should remember that royalty have never been universally beloved.

This symbolism of the eagle was already clearly established in the ancient Mesopotamian poem of Etana, possibly the first ruler ever to have his story written down. It began as an eagle and serpent swore an oath of friendship to one another before Shamish, the god of the sun. The eagle lived in the top of a tree and the serpent at its base, and, for a time, they and their young would share every kill. One day the eagle ate the young of the serpent, who then burrowed in the carcass of a bull. As soon as the eagle approached to eat, the serpent bit him, cut his wings, and threw him in a pit to die of hunger and thirst. Shamish sent the hero Etana to rescue and nurse the eagle, who eventually became his guide. Etana mounted on the back of the eagle to fly up to the heavens, and ask Ishtar, a goddess of fertility, for the plant of birth so that he might have a son. The last sections of the manuscript are fragmentary, but Etana apparently did attain his goal and founded the first Sumerian dynasty.

The Greeks later retold the story of Etana, the serpent, and the eagle as an Aesopian fable of two quarreling animals—"The Eagle and the Fox." The eagle violated its friendship with the fox by eating the fox's cubs;

the fox then set fire to the eagle's tree in revenge. The story of Etana may well have influenced the Greek myth of Ganymede, a young man who was abducted by Zeus in the form of an eagle, so that he might serve on Olympus as cupbearer of the gods. Eagles, are, however, truly capable of carrying off an infant or small child, and perhaps the story goes back to such a tragic incident.

The eagle was sacred to the Greek Zeus. It was sent by the god of thunder to eat the liver of the disobedient titan Prometheus each day, who lay chained to a rock in the Caucus Mountains. The liver would grow back during the night; this continued until the eagle was finally slain by an arrow shot by Hercules. The Roman standard was an eagle, and people conquered by Rome also often adopted the symbol.

The eagle is the initial inspiration for a huge range of mythological figures. The double-headed eagle first appears on Hittite reliefs in Mesopotamia. From there it spread to the Byzantine Empire, to eventually become a symbol of both the Holy Roman Empire and Russia. The Assyro-Babylonian epic poem "Anzu" told of the lion-headed eagle, so powerful that it could cause whirlwinds simply by flapping its wings. It once stole the Tablets of Destiny from Enlil, the god of the sky, and briefly ruled the world. Mysterious figures, sometimes known as "demon-griffins," were carved on palace walls of the Assyrian King Assyrnasirpal II. They had the bodies of men but the heads and wings of eagles, and they held up a pinecone in one hand, perhaps to enact a fertility rite

The lion shared a solar association with the eagle, and their features were often blended. Perhaps related to the lion-headed eagle, or imdugud, is the griffin, which has the face of an eagle, the body of a lion, and, sometimes at least, wings. The griffin first appeared in Mesopotamian art, but it quickly spread to Greece and beyond. Herodotus believed it lived in the mountains of India, where it made a nest of gold. Dante placed a griffin in Paradise, where it drew the chariot of the church.

Also closely related to the griffin was the Hindu Garuda, the King of Birds and the mount of Vishnu. He had the wings and beak of an eagle, and the rest of his body was human, but his vast form could darken the sky. Also inspired largely by the eagle were several other huge birds of legend such as the Arabian roc and the Persian simorgh.

In Christianity, the eagle became the symbol of Saint John the Evangelist, and is always depicted on the ground by his side. According *The Golden Legend*, written by Jacobus de Voragine in the late thirteenth century CE, this is because John once said, ". . . the eagle . . . flies higher than any other bird and looks straight into the sun, yet by its nature must come down again; and the human spirit, after it rests awhile from contemplation, is refreshed and return more ardently to heavenly thoughts." But, like an eagle, John soars straight to the mystical heights at the start of his gospel: "In the beginning was the Word . . ."

Medieval bestiaries reported that, when an eagle grew old, it would first find a fountain. Then it would fly directly into the sun until its wings were singed and it fell into the waters. After repeating this three times, the eagle would be once again filled with youthful vigor, much like Christ who rose from the dead on the third day after his burial. It also resembled the legendary phoenix, which would immolate itself and be reborn from the ashes.

One of the very few literary works in which eagles are viewed not with awe but tenderness is "The Parliament of Fowles" by Geoffrey Chaucer, written in the late fourteenth century. On Saint Valentine's Day, when the birds chose their mates, the birds gathered at the temple of Venus. Several birds paid court to the lovely female eagle that sat in the hand of the goddess. When they had all set forth their claims, Nature ruled that the female eagle herself should make the choice, thus upholding love over politics. Lords and princesses, after all, are still human beings, just as even eagles are birds.

In many ways, the Native American view of the eagle was surprisingly similar to that of Europeans. The Plains Indians, most especially, admired the strength of the eagle and associated it with the sun. The feathers of an eagle represented solar rays, and they were used on headdresses, shields, and dress to indicate skill in war or hunting. The Indians also stylized the eagle into a mythical creature—the thunderbird. The beating of its wings causes thunder, while its beak is like lightning. In Aztec religion, the golden eagle represents the sun god, Huitzilopochtli, and an eagle holding a serpent is depicted today on the Mexican coat of arms.

Depicted without the convention of heraldry, this eagle seems fiercer but more bestial.

(Illustration of bald eagle by John James Audubon, from Birds of America, c. 1828)

The eagle is a bit like a singer or actor who, after achieving great popular success, finds himself dominated by his own public image. People have trouble comprehending that the eagle, so mighty in legend, can be very vulncrable in reality. This creature has been so prominent in symbolism over the millennia that people even have trouble thinking of it as a genuine animal, and the cultural significance of the eagles seem to provide them with little protection. In countries such as the United States and Germany, eagles have become endangered despite being national emblems. Though legally protected, they are threatened by habitat loss and pesticides, though, due largely to strengthened environmental regulations, eagles are coming back in many regions.

RHINOCEROS

Ants and birds trace pattern in the dirt, but these creatures,
Armageddon in their shoulders, slip out of sight . . .

—HAROLD FARMER, "Rhinoceros"

Sometimes the legend and symbolism surrounding an animal becomes so elaborate that the creature itself is completely overshadowed, and such is the case with the rhinoceros. The creature is not one of the most important animals in myth or legend in any country. It has rarely been worshipped in temples or celebrated in epic poems. Nevertheless, sightings of the rhinoceros probably began and sustained the cult of the unicorn, which eventually also incorporated features of the horse, ass, goat, and narwhal. The irony is perhaps best illustrated in several Medieval treasuries of Europe, where the horn of a narwhal was kept as a relic of a unicorn. Alongside that horn was often that of a rhinoceros, which was believed to be a claw of a griffin. If we count the unicorn as a rhinoceros, the latter becomes one of the most important cult animals in the world.

The lore of the rhino/unicorn has a fascinating but very tangled history from the start, and there are possible depictions of it going back to prehistoric cave paintings. The first description, however, comes from the Greek Ctesias, who was the physician to the king of Persia around the start of the fourth century BCE. He considered the animal to be a giant wild ass, and nothing in nature matched his description. What suggests a rhinoceros, however, is his mention of the horn being used by Indians as a goblet as well as an antidote for poison. Rhinoceros horns have also been used for drinking, and folk medicine continues to attribute to them great potency as both a medicine and an aphrodisiac.

The description of a unicorn or "monoceros" by Pliny the Elder in the first century CE is more clearly suggestive of a rhinoceros:

> . . . the wildest animal (in India) is the monoceros, whose body is like a horse but which has the head of a stag, elephant's feet and a wild boar's tail. It utters a deep, growling sound, and a black horn, two cubits long, protrudes from the center of its forehead. It is said that the animal cannot be captured alive.

Actual rhinoceroses had appeared in triumphal processions in Rome, and Pliny may have seen the animal but failed to connect it with the accounts from travelers.

From this point on, the lore of the unicorn became ever more elaborate and romantic in Europe, and its depiction became refined and delicate. Medieval people believed that the unicorn could never be subdued

MAMMALIA ORDER BRUTA.

Rhinoceros unicornis. One horned Rhinoceros.

Illustrations of rhinos from a nineteenth-century book of natural history.

by force yet would lay its head in the lap of a virgin and allow itself to be captured. The rhinoceros, meanwhile, was usually known only from confused reports by travelers to exotic lands and, when mentioned at all, usually seemed diabolic by virtue of its brute power. When Marco Polo saw a Sumatran rhinoceros on his voyage to China, he was still able to connect it with the fabled unicorn. "All in all," he wrote, "they are nasty creatures, they always carry their piglike heads to the ground, like to wallow in the mud, and are not in the least like the unicorns of which our stories speak in Europe. Can an animal of their race feel at ease in the lap of a virgin?"

One feature of the unicorn, however, which observation did not seem to contradict was its reputation for near invincibility. In 1517, King Manuel I of Portugal brought a rhinoceros to Lisbon, the first one in Europe since Roman times. As an experiment, the King set the rhinoceros against an elephant on a street, and the pachyderm sought refuge by crashing through the iron bars of a large window. Not very long afterwards, Duke Alessandro de Medici had a rhinoceros engraved on his armor with the motto, "I make war to win."

There are many other versions of the unicorn throughout Eurasia and beyond. The oldest may be the Ky-lin, which emerged from the Yellow River before the Emperor Fu Hsi around the start of the third millennium BCE. There is also the kirin or "Japanese unicorn," which is known for its intense gaze and ability to divine whether an accused criminal is guilty or innocent. In those areas where the rhinoceros is known, it is usually at least vaguely associated with these legendary cousins.

Belief in the magical, medicinal, and aphrodisiac powers of the rhinoceros horn remains strong in the twenty-first century. Today, as many species of rhinoceros approach extinction, governments struggle with only limited success to prevent poachers from killing the animals for their horns, even though scientists affirm that these actually show no medicinal, aphrodisiac, or magical qualities. As is so often the case, research alone seems to command less credence than folkloric tradition.

Epilogue:
What is a Human Being?

nd what is a human being? There are so many answers to that question; I could write an entire book about them. But, come to think of it, I just did, and you are reading that very book now. Every section is a possible answer to the question. Human beings are animals, but not just one kind of animal. We are spiders, elephants, ravens, octopuses, butterflies . . . In writing about animals, people inevitably begin to merge human and bestial identities, anthropomorphically projecting our qualities, while simultaneously laying claim to theirs. That is why descriptions of animals, no matter what the aspiration of the author, begin to sound very dated after less than a century.

In most cases, one can learn at least as much about the writer and the culture of his era from such descriptions as one can about the animals. Literature about animals is relatively less self-conscious, and, for that reason, more revealing than writing directly about human beings. We are less defensive. Our egotism and paranoia, while present often enough, are less elaborately concealed. Animals always figure very prominently in the symbols that express human identity, from heraldic crests to the names of sports teams.

In the last half century or so, there has been an increased awareness of the mutual dependence between human beings and the natural environment. We poetically construct our identities as human beings largely through reciprocal relationships with animals. They provide us with essential points of reference, as well as illustrations of the qualities we may choose to emulate or avoid in ourselves. Any major change in our relationships with animals, individual or collective, reverberates profoundly in our character as human beings, in ways that go far beyond immediately pragmatic concerns. When a species becomes extinct, something perishes in the human soul as well.

This explains much about the cultural, religious, and philosophical diversity of human beings. Just about everyone loves animals, but people love different animals and in different ways. Almost all of us probably identify as individuals with some animals, but we have very different favorites. Pop sociology sometimes divides human beings into "dog people" and "cat people," but there are also snake, wolf, bird, and ladybug people. It is tempting to call this a sort of contemporary "totemism" or "shamanism," but such words sound much too glamorous and exotic. Our identification with animals is subsumed into our language and the mundane patterns of everyday life, occuring mostly on an unconscious level.

Is this necessarily an eternal part of the human condition? Perhaps not, since human identity is perpetually fluid. In time, perhaps machines could take over the role of animals in the construction of human identity, just as they have in many tasks such as guarding homes or plowing fields. In that case, we might not know what we had lost, but it would be a great deal indeed.

Selected Bibliography

Abrams, Roger D., editor. *Afro-American Folktales: Stories from Black Traditions in the New World*. New York: Pantheon, 1985.

Aelian. *On Animals* (3 vols.). Translated by A. F. Scholfield. Cambridge: Harvard University Press, 1972.

Aesop. *The Complete Fables of Aesop*. Translated by Olivia Temple and Robert Temple. New York: Penguin, 1998.

Aesop. *The Fables of Aesop*. Edited and retold with notes by Joseph Jacobs. New York: Macmillan, 1910.

Afanas'ev, Alekandr, editor. *Russian Fairy Tales*. Translated by Norbert Guterman. New York: Pantheon, 1973.

Alderton, David. *Crocodiles and Alligators of the World*. New York: Blanford, 1998.

Aldrovandi, Ulisse. *Aldrovandi on Chickens*. Translated by L. R. Lind. Norman, Ok: University of Oklahoma Press, 1600/2012.

Alexander of Neckam. *De Naturis Rerum*. E. Thomas Wright. London: Longman, Green, Longman, Roberts & Green, 1963.

Andersen, Hans Christian. "The Ugly Duckling." *Fairy Tales and Stories*. Translated by H. W. Dulcken. New York: Hurst, ca. 1900, 121-129.

Anderson, Virginia deJohn. *Creatures of Empire: How Domestic Animals Transformed Early America*. New York: Oxford University Press, 2004.

Andrews, Munya. "Jillinya: Great Mother of the Kimberly." In *Goddesses in World Culture*, edited by Patricia Monaghan. Santa Barbara, CA: Praeger, 2011, vol. 3. 1-12.

Anonymous, "The Worm and the Toothache." *The Ancient Near East: An Anthology of Texts and Pictures*. Edited by Pritchard, James B. Translated by W. F. Albright *et al*. Princeton, NY: Princeton University Press, 1958, 70.

Anonymous. "Children in the Wood." *In The Book of British Ballads*. Edited by S. C. Hall. London: George Routledge and Sons, 1879, 13-17.

Anonymous. *Sir Gawain and the Green Knight*. Translated by Marie Boroff. New York: W. W. Norton, 2009/ca. 1400.

Apollodorus. *The Library of Greek Mythology*. Translated by Robin Hard. New York: Oxford University Press, 1997.

Apuleius. *The Golden Ass*. Translated by E. J. Kennedy. New York: Penguin, 1999/ca. 175 CE.

Aristophanes. *The Birds*. Translated by Alan H. Sommerstein. New York: Dover, 1999.

Aristotle. *The Generation of Animals* (3 vols.). Translated by A. L. Peck. Cambridge, Mass.: Harvard U P, 1953.

Arnold, Dorothea. *An Egyptian Bestiary*. New York: Metropolitan Museum of Art, 1995.

Atchity, Kenneth J., editor and translator *The Classical Greek Reader*. New York: Oxford University Press, 1996.

Attar, Farid Ud-Din. *The Conference of Birds*. Translated by Afkham Darbandi and Dick Davis. New York: Penguin, 1984.

Attenborough, David. *The First Eden: The Mediterranean World*. Boston: Little, Brown, 1987.

Auguet, Roland. *Cruelty and Civilization: The Roman Games*. New York: Barnes & Noble, 1972.

Ausubel, Nathan. *A Treasury of Jewish Folklore*. New York: Crown, 1948.

Avianus. *The Fables of Avianus*. Translated by David R. Slavitt. Baltimore: Johns Hopkins University Press, 1993.

Ball, Philip. *The Devil's Doctor: Paracelsus and the World of Renaissance Magic and Science*. New York: Farrar, Straus and Giroux, 2006.

Barber, Lynn. *The Heyday of Natural History: 1820-1870*. Garden City, NY: Doubleday, 1980.

Baring-Gould, Sabine. *Curious Myths of the Middle Ages*. London: Longman's Green, 1892.

Baring-Gould, William S. and Ceil. *The Annotated Mother Goose*. New York: Bramhall House, 1962.

Barrie, James M. *Peter Pan*. New York: Barnes & Noble Books, 1995/1904.

Beck, Horace. *Folklore of the Sea*. Mystic, CT: 1973.

Bede (Beda Vererabilis). *Ecclesiastical History of the English Nation*. Translated by Stevens, et al. New York: Dutton, 1975.

Beer, Rüdiger Robert. *Unicorn: Myth and Reality*. Translated by Charles M. Stern. New York: Van Nostrand Reinhold, 1972.

Bergman, Charles. *Orion's Legacy: A Cultural History of Man as Hunter*. New York: Dutton, 1996.

Berlin, Isaiah. *The Hedgehog and the Fox: An Essay on Tolstoy's View of History*. New York: Simon & Schuster, 1993.

Bewick, Thomas. *A General History of Quadrupeds*. Leiscester: Winward, 1980/1790).

Bierhorst, John. *Mythology of the Lenape*. Tucson: U. of Arizona Press, 1995.

Black, Jeremy and Anthony Green. *Gods, Demons and Symbols of Ancient Mesopotamia: An Illustrated Dictionary*. Austin: U. of Texas Press, 1992.

Blake, William. "The Sick Rose. "*The Norton Anthology of English Literature* (2 vols.). Edited by .H. Abrams. Fifth Edition. New York: W. W. Norton, 1986, vol. 2, 40.

Blake, William. "The Tyger." *The Oxford Book of Animal Poems*. Edited by Michael Harrison and Christopher Stuart-Clark. New York: Oxford University Press, 1992.

Boia, Lucian. *Entre l'ange et la bête: Le mythe de l'homme différent de   l'Antiquité à nos jours*. Paris: Plon, 1995.

Botkin, B. A., editor. *A Treasury of American Folklore: Stories, Ballads, and Traditions of the American People*. New York: Crown Publishers, 1944.

Briggs, Katharine. *Nine Lives: The Folklore of Cats*. New York: Dorset Press, 1980.

Brinton, Daniel G. *Myths of the New World*. 3rd edition. New York: Leypoldt & Holt, 1896.

Brown, Joseph Epes. *Animals of the Soul: Sacred Animals of the Oglala Sioux*. Rockport, Mass.: Element, 1992.

Bruchac, Joseph. *Native Plant Stories*. Golden, CO.: Fulcrum, 1995.

Buhs, Joshua Blu. *Bigfoot*. Chicago: University of Chicago Press, 2009.

Burnett, Frances Hodgson. *The Secret Garden*. New York: Barnes and Noble Books, 1990.

Burt, Jonathan. *Rat*. London: Reaktion Books, 2006.

Bushnaq, Inea, editor and translator. *Arab Folktales*. New York: Pantheon, 1986.

Calvino, Italo. *Fiabe Italiano: Raccolte e transcitte da Italo Calivino* (2 vols.). Milan: Oscar Mondadori, 1986.

Campbell, Joseph. *The Hero with a Thousand Faces*. Princeton, NJ: Princeton University Press, 1990.

Campbell, Joseph. Vol. Part 2, *Mythologies of the Great Hunt, Historical Atlas of World Mythology New York*: Harper & Row, 1988.

Cao Xueqin, and Gao E. *The Story of the Stone, Also Known as the Dream of the Red Chamber*. Translated by John Minford. 5 vols. New York: Penguin, 1986/1792.

Caras, Roger A. *A Perfect Harmony: The Intertwining Lives of Animals and Humans throughout History*. New York: Simon and Schuster, 1996.

Carlson, Rev. Gregory I. "Horace's and Today's Town and Country Mouse." *Bestia*, vol. 4 (May 1992), 87-112.

Carpenter, Francis. *Tales of a Korean Grandmother*. Rutland, VT: Charles E. Tuttle, 1973.

Catullus, "*Lugete, o Veneres.*" *Catullus, Tibullus, Pervigilim Veneris*. Edited and translated by C. P. Goold. Cambridge, Mass.: Harvard University Press, 1962, 4-5.

Cavalieri, Paola, and Peter Singer, editors. *The Great Ape Project: Equality beyond Humanity*. New York: Saint Martin's Press, 1995.

Charbonneau-Lassay, Louis. *The Bestiary of Christ*, Edited and translated by D. M. Dooling. New York: Parabola Books, 1991.

Chaucer, Geoffrey. "The Parliament of Fowls." *The Works of Geoffrey Chaucer*. Second edition. Edited by F. N. Robinson. Boston: Houghton Mifflin, 1961, 309-318.

Chaucer, Geoffrey. *The Canterbury Tales*. New York: Penguin, 2005.

Christie, Anthony. *Chinese Mythology*. New York: Barnes & Noble Books, 1996.

Cicero. *The Nature of the Gods*. Translated by Horace C. P. McGregor. New York: Penguin, 1972.

Clutton-Brock, Juliet . *Horse Power: A History of the Horse and the Donkey in Human Societies*. Cambridge, Mass.: Harvard University Press, 1992.

Cogger, Hardold G. et al, editors. *Encyclopedia of Animals*. San Francisco: Fog City Press, 1993.

Cohen, Daniel. *The Encyclopedia of Monsters*. New York: Dorset Press, 1982.

Collodi, Carlo. *Pinocchio*. Translated by E. Harden. New York: Puffin Books, 2009/1882.

Comfort, David. *The First Pet History of the World. New York: Fireside, 1994.*

Coomaraswamy, Ananda K., and Sister Nivedita. *Myths of the Hindus and Buddhists*. New York: Dover, 1967.

Cooper, J. C. *Symbolic and Mythological Animals*. London: Harper-Collins/ The Aquarian Press, 1992.

Corbey, Raymond, and Bert Theunissen, editors. *Ape, Man, and Apeman: Changing Views since 1900*. Leiden: Leiden University Press, 1995.

Courlander, Harold. *A Treasury of African Folklore: The Oral Literature, Traditions, Myths, Legends, Epics, Tales, Recollections, Wisdom, Sayings, and Humor of Africa*. New York: Marlowe and Co., 1996.

Courtney, Nicholas. *The Tiger: Symbol of Freedom*. New York: Quartet Books, 1980.

Courtright, Paul B. Ganeša: *Lord of Obstacles, Lord of Beginnings*. New York: Oxford University Press, 1985.

Crane, Hart. "Voyages." In *The Oxford Book of American Poetry*, edited by David Lehman. New York: Oxford University Press, 1779/1926, 434-435.

Cummins, John. *The Hound and the Hawk: The Art of Medieval Hunting*. London: Phoenix Press, 1988.

D'Aulnoy, Marie Catherine. "The White Cat." Translated by Minnie Wright. In *The Blue Fairy Book*. Edited by Andrew Lang. New York: Dover, 1965, 157-173.

Dähnhardt, Oskar. *Naturgeschichtliche Volksmärchen*, vol. 1. Leipzig: B. G. Teubner, 1909.

Dale, Rodney. *Cats in Boots: A Celebration of Cat Illustration through the Ages*. New York: Harry N. Abrams, 1997.

Dalley, Stephanie, editor. *Myths from Mesopotamia: Creation, The Flood, Gilgamesh and Others*. New York: Oxford University Press, 1993.

Dasent, George. "East of the Sun and West of the Moon." *In Scandinavian Folk and Fairy Tales*. Edited by Claire Booss. New York: Avenel Books, 1984, 63-70.

Davies, Malcolm and Jeyaraney Kathirithamby. *Greek Insects*. New York: Oxford University Press, 1986.

Davies, Malcolm and Jeyaraney Kathirithamby. *Greek Insects*. New York: Oxford University Press, 1986.

Davis, Courtney and Dennis O'Neil. *Celtic Beasts: Animal Motifs and Zoomorphic Design in Celtic Art*. London: Blandford, 1999.

Degraaff, Robert M. *The Book of the Toad: A Natural and Magical History of Toad-Human Relations*. Rochester, Vt.: Park Street Press, 1991.

Dekkers, Midas. *Dearest Pet: On Bestiality*. Translated by Paul Vincent. New York: Verso, 2000.

Delort, Robert. *Les animaux ont une histoire*. Paris: Éditions du Seuil, 1984.

Delort, Robert. *The Life and Lore of the Elephant*. Translated by I. Mark Paris. New York: Harry N. Abrams, 1992.

Dent, Anthony. *Donkey: The Story of the Ass from East to West*. Washington, D.C.: George G. Harrap, 1972.

Dent, Anthony. *The Horse: Through Fifty Centuries of Civilization*. New York: Holt, Rinehart and Winston, 1974.

Derrida, Jacques. *"Che cos' è la poesia?" Points: Interviews, 1974-1994*. Translated by Peggy Kampuf et al. Edited by Elizabeth Weber. Stanford: Stanford University Press, 1992, 288-300.

Derrida, Jacques. *The Animal That Therefore I Am*. Translated by David Willis. New York: Fordham UP, 2008.

Digard, Jean-Pierre. *L'homme et les animaux domestiques*. Paris: Fayard, 1990.

Dobie, J. Frank. *The Voice of Coyote*. Lincoln, Nebraska: U. of Nebraska Press, 1961.

Donne, John. "The Flea." *Animal Poems*. Edited by John Hollander. New York: Knopf, 1994, 127-128.

Dostoyevski, Fyodor. *Crime and Punishment*. Translated by David Magarshak. New York: Greenwich House, 1982/1865.

Douglas, Mary. *Purity and Danger: An Analysis of the Concepts of Pollution and Taboo*. New York: Routledge, 2002/1946.

Douglas, Norman. *Birds and Beasts of the Greek Anthology*. New York: Jonathan Cape and Harrison Smith, 1929.

Eckholm, Eric. "China's Little Gladiators, Fearsome in the Ring." *The New York Times*, Oct. 4, 2000, 4.

Eco, Umberto, editor *History of Beauty*. Translated by Alastair McEwen. New York: Rizzoli, 2004.

Edwards, Jonathan. "Sinners in the Hands of an Angry God." In *The Sermons of Jonathan Edwards: A Reader*. Edited by Wilson H. Kimnach, Kenneth P. Minkena, and Douglas A. Sweeney. New Haven: Yale University Press, 1999/1734, 49-65.

Eisler, Colin. *Dürer's Animals*. Washington, DC: Smithsonian Institution Press, 1991.

Eliade, Mircea, *History of Religious Ideas*, translator Willard R. Trask, 3 vols. Chicago: University of Chicago Press, 1998.

Eliade, Mircea. *Shamanism: Archaic Techniques of Ecstasy*. Princeton: Princeton University Press, 1974.

Elston, Catherine Feher. *Ravensong: A Natural and Fabulous History of Ravens and Crows*. Flagstaff, AZ: Northland, 1991.

Erdoes, Richard and Alfonzo Ortiz. *American Indian Trickster Tales*. New York: Penguin, 1988.

Eschenbach, Wolfram von. *Parzifal*. Translated by A. T. Hatto. New York: Penguin, 1980.

Estés, Clarissa Pinkola. *Women Who Run with the Wolves: Myths and Stories of the Wild Woman Archetype*. New York: Ballantine Books, 1992.

Evans, E. P. *The Criminal Prosecution and Capital Punishment of Animals: The Lost History of Europe's Animal Trials*. Boston: Faber and Faber, 1988.

Fabre, Henri, *Fabre's Book of Insects*. Translated by Mrs. Rodolph Stawell. New York: Tudor, 1935.

Fabre-Vassas, Claudine. *The Singular Beast: Jews, Christians, and the Pig*. Translated by Carol Volk. New York: Columbia University Press, 1997.

Faulkner, William. "The Bear." In *Go Down Moses*. New York: Vintage Books, 1970, 181-316.

Flores, Nona C. "Effigies Amicitiae . . . Veritatas Inimicitiae": Antifeminism in the Iconograhy of the Woman-headed Serpent in Medieval and Renaissance Art and Literature." *Animals in the Middle Ages: A Book of Essays*. Edited by Nona C. Flores. New York: Garland, 1996, 167-196.

Fontenay, Elizabeth de. *Le silence des bêtes: La philosophie à l'épreuve de l'animalité*. Paris: Fayard, 1998.

Gantz, Jeffrey, editor and translator. *Early Irish Myths and Sagas*. New York: Penguin Books, 1982.

Gibson, Claire. *Signs and Symbols: An Illustrated Guide to their Meaning*. New York: Barnes & Noble, 1996.

Giles, Herbert A. Translated by *Strange Stories from a Chinese Studio*. New York: Boni and Liveright, 1926,

Gill, Sam D. and Irene F. Sullivan. *Native American Mythology*. New York: Oxford University Press, 1992.

Gimbutas, Marija. *The Goddesses and Gods of Old Europe: Myths and Cult Images*. New York: U. of California Press, 1992.

Giorgetti, Anna. *Ducks: Art, Legend, History*. Translated by Helena Ramsay. Boston: Little, Brown and Company/Bullfinch Press, 1992.

Girourd, Mark. *The Return to Camelot: Chivalry and the English Gentleman*. New Haven, CT: Yale University Press, 1985.

Glassie, Henry, editor. *Irish Folk Tales*. New York: Pantheon, 1985.

Glickman, Stephen. "The Spotted Hyena from Aristotle to the Lion King: Reputation is Everything." *Social Research*, vol. 62, # 3 (fall 1995): p. 501-539.

Goethe, Johann Wolfgang von. "Blissful Longing." Translated by Albert Bloch. *Anthology of German Poetry through the Nineteenth Century*. Edited by Alexander Gode and Frederick Ungar. New York: Frederick Ungar, 1964, p 102-103.

Goldsmith, Oliver. *History of Animated Nature (4 vols.)*. Edinburgh: Smith, Elder/T. Tegg, 1838.

Gordon, Edmund I. "Sumerian Animal Proverbs: 'Collection Five.'" *Journal of Cuneiform Studies* 12 (1958): 1-6.

Gotfredsen, Lise. *The Unicorn*. New York: Abbeville Press, 1999.

Graham, Alistair. *Eyelids of Morning: The Mingled Destinies of Crocodiles and Men*. New York: A & W Visual Library, 1973.

Graham, Lanier. "Goddess Androgyne: Coatlicue of the Aztecs." In *Goddesses in World Culture*, edited by Patricia Monaghan. Santa Barbara, CA: Praeger, 2011, vol. 3, 73-84.

Grahame, Kenneth. *The Wind and the Willows*. Hollywood, FL: Simon & Brown, 1012/1907.

Grantz, Jeffrey, translator. *The Mabinogion*. New York: Dorset Press, 1976.

Grass, Günter. *The Call of the Toad*. Translated by Ralph Manheim. New York: Harcourt, Brace & Janovich, 1993.

Grimm, Jacob and Wilhelm. *The Complete Fairy Tales of the Brothers Grimm*. Translated by Jack Zipes. New York: Bantam Books, 1987.

Grimm, Jacob and Wilhelm. *The German Legends of the Brothers Grimm* (2 vols.) Edited and translated by Donald Ward. Philadelphia: Institute for the Study of Human Issues, 1981.

Gubernatis, Angelo De. *Zoological Mythology or The Legends of Animals* (2 vols.). Chicago: Singing Tree Press, 1968.

Hall, James. *Illustrated Dictionary of Symbols in Eastern and Western Art*. New York: HarperCollins, 1996.

Hall, S. C., editor. *The Book of British Ballads*. London: George Routledge and Sons, 1879, 373-374.

Ha-Nakdan, Berechiah. *Fables of a Jewish Aesop*. Translated by Moses Hadas. New York: Comumbia University Press, 1967.

Haraway, Donna J. *When Species Meet*. Minneapolis: University of Minnesota Press, 2008.

Haraway, Donna. *Primate Visions: Gender, Race, and Nature in the World of Modern Science*. New York: Routledge, 1990.

Hardy, Thomas. "The Darkling Thrush." *The Norton Anthology of English Literature*. vol. 2. 5th Edition. Edited by M. H. Abrams. New York: W. W. Norton, 1986, 1743-1744.

Harris, Joel Chandler. *Uncle Remus: His Stories and His Sayings*. New York: A. Appleton, 1928.

Hausman, Gerald. *Meditations with Animals: A Native American Bestiary*. Santa Fe: Bear, 1986.

Hawley, Fred. "The Moral and Conceptual Universe of Cockfighters: Symbolism and Rationalization." *Society and Animals*. Vol. 1, # 2 (1993), 159-168.

Hearn, Lafcadio. *Kwaidan: Stories and Studies of Strange Things*. Rutland, VT: Charles E. Tuttle, 1971.

Hearn, Lofcadio. *Japanese Fairy Tales*. Mount Vernon, NY: Peter Pauper Press, 1936.

Hell, Betrand. "Enraged Hunters: The Domain of the Wild in Northwestern Europe." In *Nature and Society: Anthropological Perspectives*, edited by Philippe Descola and Gísli Pálsson, 214-55. New York: Routledge, 1996.

Hendrickson, Robert. *More Cunning than Man: A Social History of Rats and Men*. New York: Dorset Press, 1983.

Henish, Bridget Ann. *Fast and Feast: Food in Medieval Society*. University Park, Penn.: Pennsylvania University Press, 1994.

Herodotus. *Herodotus* (4 vols.). Translated by A. D. Godley. New York: G. P. Putnam's Sons, 1926.

Hesiod. *Theogony/Works and Days*. Translated by M. L. West. New York: Oxford University Press, 1988.

Hillyard, Paul. *The Book of the Spider: From Arachnophobia to the Love of Spiders*. New York: Random House, 1994.

Hine, Daryl, translator *The Homeric Hymns and The Battle of the Frogs and the Mice*. New York: Anthenium, 1972.

Hoffmann, E. T. A. *The Life and Opinions of Kater Murr*. In *Selected Writings of E. T. A. Hoffmann* (2 vols.). Edited and translated by Leonard J. Kent and Elizabeth C. Knight. Chicago: University of Chicago Press, 1969/1820.

Hole, Christina, E. Radford and M. A. Radford. *The Encyclopedia of Superstitions*. New York: Barnes & Noble Books, 1996.

Houlihan, Patrick F. *The Animal World of the Pharaohs*. New York: Thames and Hudson, 1996.

Howell, Signe. "Nature in Culture or Culture in Nature? Chewong Ideas of 'Humans' and Other Species'." In *Nature and Society: Anthropological Perspectives*, edited by Philippe Descola and Gísli Pálsson, 152-68. New York: Routledge, 1996.

Hughes, D. Wyn, Edited by *Hares*. New York: Congdon & Lattès, 1981.

Hugo, Victor. *The Hunchback of Notre Dame*. Translated by Catherine Liu. New York: Modern Library Classics, 2002/1831.

Hurston, Zora Neale. *Mules and Men*. New York: HarperTrade, 1990.

Hutton, Ronald. *The Stations of the Sun: A History of the Ritual Year in Britain*. New York: Oxford University Press, 1997.

Hyland, Ann. *Equus: The Horse in the Roman World*. New Haven: Yale University Press, 1990.

Hyland, Ann. *The Medieval Warhorse from Byzantium to the Crusades*. Conshohocken, PA: Combined Books, 1994.

Ions, Veronica. *Egyptian Mythology*. New York: Peter Bedrick Books, 1982.

Jacobs, Joseph, editor "Dick Whittington and His Cat." In *English Fairy Tales*. New York: Dover, 1967/1890, 167-178.

Jameson, R. D. "Cinderella in China." *Cinderella: A Casebook*. Edited by Alan Dundes. Madison: U. of Wisconsin Press, 1988, 71-97.

Jiménez, Juan Ramón. *Platero and I*. Translated by Eloise Roach. Austin: U. of Texas Press, 1983.

Johnson, Allison. *Islands in the Sound: Wildlife in the Hebrides*. London: Victor Gollancz, 1989.

Johnson, Buffie. *Lady of the Beasts: Ancient Images of the Goddess and Her Sacred Animals*. San Francisco: Harper & Row, 1981.

Johnson-Davies, Denys, translator. *The Island of Animals*. Austin: U. of Texas Press, 1994.

Johnston, Johanna. *The Fabulous Fox: An Anthology of Fact and Fiction*. New York: Dodd, Mead & co., 1979.

Jones, Alison. *Larousse Dictionary of World Folklore*. New York: Larousse, 1996.

Jones, Gwyn, translator, "King Hrolf and His Champions." *In Erik the Red and Other Icelandic Sagas*. New York: Oxford University Press, 1991, 221-318.

Joyce, James. *Portrait of the Artist as a Young Man*. New York: Barnes & Noble Books, 1999/1916.

Joyce, P. W., editor *Old Celtic Romances: Tales from Irish Mythology*. New York: Devin-Adair, 1962.

Kafka, Franz. "Metamorphosis." *The Complete Stories*. Translated by Willa and Edwin Muir et al. New York: Schocken, 1971/1915, 89-139.

Kipling, Rudyard. *The Two Jungle Books*. Garden City, NJ: Sun Dial Press, 1895.

Klingender, Francis. *Animals in art and thought till the end of the Middle Ages*. Translated by Evelyn Antal and John Harthan. Cambridge, Mass.: MIT University Press, 1971.

Knappert, Jan. *African Mythology: An Encyclopedia of Myth and Legend*. London: Diamond Books, 1995.

La Fontaine, Jean. "The Pigeon and the Ant." *Selected Fables*. Translated by James Michie. New York: Viking, 1979, 49.

Laduke, Winona. *All Our Relations: Native Struggles for Land and Life*. Cambridge, MA: South End Press, 1999.

Lady Wilde (Speranza). *Ancient Legends, Mystic Charms, and Superstitions of Ireland: With Sketches of the Irish Past*. Galway: O'Gorman, 1888/1971.

Lambert, W. G. *Babylonian Wisdom Literature*, Oxford: Clarendon Press, 1960.

Larrington, Caroline, Translated by *The Poetic Edda*. New York: Oxford University Press, 1996.

Lawrence, Elizabeth Atwood. *Hunting the Wren: Transformation of Bird to Symbol.* Knoxville: University of Tennessee Press, 1997.

Leach, Maria. *God had a Dog: Folklore of the Dog.* New Brunswick, NJ: Rutgers University Press, 1961

Leopold, Aldo. "Thinking like a Mountain." *A Sand County Almanac: and selected sketches from here and there.* New York: Oxford University Press, 1949, 129-133.

LeRoy, Marcel. "Lettres sur les Animaux." *Variétés Litteréraires Recueil des Pieces tant originales que trauites, concernant las Philosophie, la Littérature et les Arts.* Paris: Lacombe, 1768, vol. 3, p. 1-173.

Lewinsohn, Richard. *Animals, Men and Myths: An Informative and Entertaining History of Man and the Animals Around Him. New York: Harper & Brothers,* 1954.

Lewis, C. S. *The Lion, the Witch, and the Wardrobe.* New York: HarperCollins, 1994/1950.

London, H. Stanford et al. *The Queen's Beasts: An Account with New Drawings of the Heraldic Animals which Stood at the Entrance to Westminster Abbey on the Occasion of the Coronation of Her Majesty Queen Elizabeth II.* London: Newman Neame, 1953.

Lopez, Barry Holstun. *Of Wolves and Men.* New York: Charles Scribner's Sons, 1978.

Maeterlink, Maurice. *Life of the Bee.* Translated by Alfred Sutro. London: George Allen, 1908.

Mandeville, Bernard. *The Fable of the Bees: or Private Vices, Publick Benefits* (2 vols.): Oxford, Clarendon Press, 1924.

Marigny, Jean. *Vampires: Restless Creatures of the Night.* Translated by Lory Frankel. New York: Harry N. Abrams, 1994.

Matamonasa-Bennett, Arieahn. "White Buffalo Calf Woman (Pte-San Win-Yan): The First and Second Coming." In *Goddesses in World Culture,* edited by Patricia Monaghan. Santa Barbara, CA: Praeger, 2011, vol. 3, 161-175.

Matarasso, Pualine Maud, editor *The Quest of the Holy Grail.* New York: Penguin, 1969.

McCullough, Helen Craig, translator. *The Tale of the Heike.* Stanford: Stanford University Press, 1988.

Medlin, Faith. *Centuries of Owls in Art and the Written Word.* Norwalk, CT: Silvermine Publishers, 1968.

Melegros. "The Mosquito." *Poems from the Greek Anthology.* Edited and translated by Dudley Fitts. New York: New Directions, 1956, 26.

Melville, Herman. *Moby-Dick, Or The Whale*. New York: Penguin, 1992/1851.

Menache, Sophia. "Dogs: God's Worst Enemies?" *Society and Animals. Vol. 5, # 1 (1997): 23-44.*

Milton, John. *Paradise Lost and Paradise Regained*. New York: Penguin, 1976.

Mohammad (attributed to). *The Koran, with a Parallel Arabic Text*. Translated by N. J. Dawood. New York: Penguin, 2000.

Montaigne, Michel de. "Apology for Raymond Sebond." In The Complete Essays of Montaigne (2 vols.). Translated by Donald M. Frame. Stanford: Stanford University Press, 1959, vol. 1, 428-561.

Mourning Dove (Humishuma). *Coyote Stories*. Lincoln, Nebraska: U. of Nebraska Press, 1990.

Mowat, Farley. *Never Cry Wolf*. Boston: Little, Brown, 1963.

Mullett, G. M. *Spider Woman Stories*. Tucson: University of Arizona Press, 1979.

Mundkur, Balaji. *The Cult of the Serpent: An Interdisciplinary Survey of its Manifestations and Origins*. Albany: SUNY Press, 1983.

Nagel, Thomas. "What is it like to be a bat?" *Moral Questions*. New York: Cambridge University Press, 1985, p. 169-180.

Netboy, Anthony. *The Salmon: Their Fight for Survival*. Boston: Houghton Mifflin, 1973.

Nigg, Joe. *The Book of Gryphons*. Cambridge, Mass.: Applewood Books, 1982.

Nigg, Joseph. *The Book of Fabulous Beasts: A Treasury of Writings from Ancient Times to the Present*. New York: Oxford University Press, 1999.

Nissenson, Marilyn, and Susan Jones. *The Ubiquitous Pig*. New York: Harry N. Abrams, 1996.

Nott, Charles Stanley. *The Flowery Kingdom*. New York: Chinese Study Group of America, 1947.

O, Meara, John J., translator. *The Voyage of Saint Brenden: Journey to the Promised Land*. Atlantic Highlands, NJ: Humanities Press, 1976.

O. Neill, J. P. *The Great New England Sea Serpent: An Account of Unknown Creatures Sighted by Many Respectable Persons between 1638 and the Present Day*. Camden, ME: Down East Books, 1999.

O'Flaherty, Wendy Doniger. *Women, Androgynes, and Other Mythical Beasts*. Chicago: U. of Chicago Press, 1980.

O'Sullivan, Patrick. *Irish Superstitions and Legends of Animals and Birds*. Dublin: Mercier Press, 1991.

Odo of Cheriton. *The Fables of Odo of Cheriton*. Translated by John C. Jacobs. New York: Syracuse University Press, 1985.

Olalquiaga, Celeste. *The Artificial Kingdom: A Treasury of Kitsch Experience*. New York: Pantheon, 1998.

Oppian. "The Loves of the Tortoise, from *Halieutica*." Translated by William Diaper. *Animal Poems*. Edited by John Hollander. New York: Knopf, 1994, 158.

Orwell, George. *Animal Farm.* New York: Harcourt, Brace and Janovich/Signet Classics, 1946.

Osborne, Harold. *South American Mythology.* New York: Hamlyn, 1975.

Ovid. *Metamorphoses.* Translated by Rolfe Humphries. Bloomington: Indiana University Press, 1955.

Parrinder, Geoffrey. *African Mythology.* London: Paul Hamlyn, 1973.

Pastoreau, Michel. *Le Cochon: Histoire D'un Cousin Mal Aimé.* Paris: Gallimard, 2009.

Pastoureau, Michael. *Les Animaux Célèbres.* Paris: Arléa, 2008.

Pastoureau, Michael. *The Bear: History of a Fallen King.* Translated by George Holoch. Cambridge, MA: Bellknap Press, 2007.

Perrault, Charles. *Perrault's Fairy Tales, with Thirty-Four Full-Page Illustrations by Gustave Doré.* Translated by A. E. Johnson. New York: Dover, 1969/1697.

Perrault, Charles. *Perrault's Fairy Tales.* Translated by A. E. Johnson. New York: Dover, 1969/1697.

Perry, Ben Edwin, translator and editor. *Babrius and Phaedrus.* Cambridge, Mass.: 1965.

Petronius Arbiter. *Petronius/ Sececa, Apocolocyntosis.* Translated by Michael Heseltine, W. H. D. Rouse, E. H. Warmington. Cambridge: Harvard University Press, 1997.

Plato. "Phaedrus." *The Collected Dialogues of Plato.* Translated by R. Hackforth, Edited by Edith Hamilton and Huntington Cairns. New York: Pantheon (Bollingen Series), 1961, 475-525.

Plato. *The Last Days of Socrates.* Translated by Hugh Tredennick and Harold Tarrant. New York: Penguin, 1993.

Pliny. *Natural History* (10 vols.). Translated by H. Rackham, W. H. S. Jones *et al.* Cambridge, Mass.: Harvard University Press, 1953.

Pliny the Elder. *Natural History: A Selection.* Edited and translated by John F. Healey. New York: Penguin, 1991.

Plutarch. "Isis and Osiris." *Plutarch's Moralia (15 vols.).* Translated by Frank Cole Babbitt et al. Cambridge, Mass.: Harvard University Press, 1962, vol. 5, 3-384.

Plutarch. *Greek Lives: A Selection of Nine Greek Lives.* Translated by Robin A. Waterfield. New York: Oxford University Press, 1999.

Plutarch. *Moralia.* Translated by Frank Cole Babbitt. 15 vols. Cambridge, MA: Harvard Up (Loeb Classic Library), 1936.

Plutarch. *Plutarch's Morals* (4 vols.). Translated by Robert Midgley *et al.* London: Thomas Bradyll, 1704.

Poe, Edgar Allan. "The Raven." *Last Flowers: The Romance Poems of Edgar Allan Poe*

Poe, Edgar Allan. *Complete Stories and Poems of Edgar Allan Poe*. New York: Doubleday, 1966.

Pollack, Andrew. " Seeking Cures, Patients Enlist Mice Stand-Ins"." *New York Times*, 2012, A1, A3.

Pollard, John. *Birds in Greek Life and Myth*. New York: Thames and Hudson, 1977.

Potter, Beatrix. *Mrs. Twiggy-Winkle*. London: Frederick Warne, 1905.

Potter, Beatrix. *The Tale of Squirrel Nutkin*. New York: Frederick Warne, 1012/1903.

Preston, Claire. *Bee*. London: Reaktion Books, 2006.

Price, A. Lindsay. *Swans of the World: In Nature, History, Myth and Art*. Tulsa, OK: Council Oak Books, 1994.

Pritchard, James B., editor *The Ancient Near East: An Anthology of Texts and Pictures*. Princeton: Princeton University Press, 1958.

Rappoport, Angelo S. *The Sea: Myths and Legends*. London: Senate, 1995.

Raspe, Rudolph Erich. *The Surprising Adventures of Baron Munchausen*. No translator given. New York: Everyman's Library, 2012/1785.

Rilke, Rainer Maria. "The Panther." Translated by Stephen Mitchell. Animal Poems. *Animal Poems*. Edited by John Hollander. New York: Everyman's Library, 1994, 103.

Robbins, Mary E. *"The Truculent Toad in the Middle Ages." Animals in the Middle Ages: A Book of Essays*. Edited by Nona C. Flores. New York: Garland, 1996.

Roob, Alexander. *Alchemy and Mysticism*. New York: Taschen, 1997.

Root, Nina J. "Victorian England's Hippomania: From the Nile to the Thames they loved Obaysch." *Natural History*. February 1993, 34-38.

Rothenberg, David. *Why Birds Sing: A Journey into the Mystery of Bird Song*. New York: Basic Books, 2005.

Rowland, Beryl. *Animals with Human Faces: A Guide to Animal Symbolism*. Knoxville: U. of Tenn. Press, 1973.

Rowling, J. K. *Harry Potter and the Sorcerer's Stone*. New York: Scholastic/Arthur A. Levine Books, 1998.

Ryan, W. F. *The Bathhouse at Midnight: Magic in Russia*. University Park, Penn.: Pennsylvania State University Press, 1999.Rybot, Doris. *It Began Before Noah*. London: Michael Joseph, 1972.

Rybot, Doris. *It Began Before Noah*. London: Michael Jospeh, 1972.

Ryder, Arthus W., editor, *The Panchatantra*. Chicago: University of Chicago Press, 1964.

Sachs, Hans. "Ursprung der Affen." In *Hans Sachsens Ausgewählte Werke* (2 vols.). Leipzig: Insel Verlag, 1945, vol. 1, 166-169.

Salten, Felix. *Bambi: A Life in the Woods*. New York: Simon and Schuster, 1928.

Santino, Jack. *All around the Year: Holidays and Celebrations in American Life*. Chicago: University of Illinois Press, 1995.

Sax, Boria. "What Is This Quintessence of Dust? The Concept of the 'Human' and Its Origins." In *The End of Anthropocentrism*, edited by Rob Boddice, pp. 21-37. London: Berg, 2011.

Sax, Boria. *Animals in the Third Reich: Pets, Scapegoats, and the Holocaust*. New York: Continuum, 2000.

Sax, Boria. Bestial Wisdom and Human Tragedy: The Genesis of the Animal Epic." *Anthrozoös*. Vol. 11, # 4 (1998), 134-141.

Sax, Boria. *City of Ravens: London, Its Tower, and Its Famous Birds*. London: Duckworth-Overlook, 2011-2012.

Sax, Boria. *The Frog King: On Legends, Fables, Fairy Tales and Anecdotes of Animals*. New York: Pace University Press, 1990.

Sax, Boria. *The Parliament of Animals: Legends and Anecdotes, 1775-1900*. New York: Pace University Press, 1990.

Sax, Boria. *The Serpent and the Swan: Animal Brides in Folklore and Literature*. Blacksburg, VA: McDonald & Woodward, 1998.

Schochet, Elijah Judah. *Animals in Jewish Tradition: Attitudes and Relationships*. New York: Ktav Publishing House, 1984.

Schrader, J. L. *A Medieval Bestiary*. New York: Metropolitan Museum of Art, 1986.

Schwarz, Marion. *A History of Dogs in the Early Americas. New Haven, Yale University Press, 1997.*

Scott, Sir Walter. *Letters on Demonology and Witchcraft. New York: J. & J. Harper, 1832.*

Seton-Thompson, Ernest. *Wild Animals I have Known*. New York: Charles Scribner's, 1990.

Sewell, Anna. *Black Beauty*. New York: Dover, 1999/1877.

Shakespeare, William. *A Midsummer Night's Dream*. New York: Bantam Classics, 1988.

Shakespeare, William. *Love's Labor Lost*. New York: Viking Penguin, 2000.

Shakespeare, William. *Romeo and Juliet*. New York: Dover, 1993.

Shakespeare, William. *Twelfth Night or, What You Will. New York: Dover, 1966,*

Sheasley, Bob. *Home to Roost: A Backyard Farmer Chases Chickens through the Ages*. New York: Thomas Dunne Books, 2008.

Shepard, Paul. *The Others: How Animals Made Us Human*. Washington, D. C.: Shearwater Books, 1996.

Shephard, Odell. *The Lore of the Unicorn*. New York: The Metropolitan Museum of Art, 1982.

Shepherd, Paul, and Barry Sanders. *The Sacred Paw: The Bear in Nature, Myth, and Literature*. New York: Viking Penguin, 1992.

Shoeki, Ando. *The Animal Court: A Political Fable from Old Japan*. Translated by Jeffrey Hunter. New York: Weatherhill, 1992.

Sillar, F. C. and R. M. Meyler, editors. *Elephants, ancient and modern*. New York: Viking Press, 1968.

Sleigh, Charlotte. *Ant*. London: Reaktion Books, 2003.

Smart, Christopher. "My Cat Jeoffry." In *Animal Poems*, edited by John Hollander. New York: Knopf, 1994/1762, 27-31.

Smith, Lacey Baldwin. *Fools, Martyrs, Traitors: The Story of Martyrdom in the Western World*. New York: Knopf, 1997.

Spiekermann, Uwe. "Das Andere Verdauen: Begegnungen Von Ernährungskulturen." In *Ernährung in Grenzsituationen*, edited by Uwe Spiekermann and Gesa U. Schönberger, Heidelberg: Springer Verlag, 2002, 89-106.

Sprenger, James, and Heinrich Kramer. *The Malleus Malificarum*. New York: Dover, 1971/1484.

Staal, James D. *The New Patterns in the Sky: Myths and Legends of the Stars*. Blacksburg, VA: McDonald & Woodward, 1988.

Staal, Julius D. W. *The New Patterns in the Sky: Myths and Legends of the Stars*. Blacksburg, VA: McDonald & Woodward, 1988.

Steedman, Amy. *Legends and Stories from Italy*. New York: G. P. Putnam's Sons, ca. 1910.

Stevens, Wallace. "Sunday Morning *Stevens/Poems*. New York: Knopf/Everyman's Library, 1993, 40-41.

Stone, Brian, translator. *The Owl and the Nightingale/ Cleanness/ St Erkenwald*. Second Edition. New York: Penguin, 1988.

Suetonius. *Suetonius* (2 vols.). Translated by J. C. Rolfe. Cambridge: Harvard University Press, 1998.

Sun, Ruth Q. *The Asian Animal Zodiac*. Boston: Castle Books, 1974.

Swift, Jonathan. "The Battle of the Books." In *The Writings of Jonathan Swift*. Edited by Robert A. Greenberg and William B. Piper. New York: Norton, 1973/1704, 373-396.

Swift, Jonathan. *Gulliver's Travels*. New York: Penguin, 1987/1726.

Taleb, Nassim Nicholas. *The Black Swan: The Impact of the Highly Improbable*. New York: Random House, 2007.

Tatar, Maria. *The Annotated Brothers Grimm*. New York: W. W. Norton, 2004.

Terry, Patricia, editor and translator. *Renard the Fox*. Boston: Northeastern University Press, 1983.

Thompson, Flora. *Lark Rise to Candleford: A Trilogy*. New York: Penguin, 1973.

Thomson, David. *The People of the Sea: A Journey in Search of the Seal Legend*. Washington, DC: Counterpoint, 2000.

Thoreau, Henry David. *Walden, and Other Writings of Henry David Thoreau*. New York: Modern Library, 1965.

Thurber, James. *Fables for Our Time and Famous Poems*. New York: Harper Colophon, 1990/1940.

Thurston, Mary Elizabeth. *The Lost History of the Canine Race: Our 15,000-Year Love Affair with Dogs.* New York: Avon Books, 1996.

Tompkins, Ptolemy. *The Monkey in Art.* Wappinger's Falls, NY: M. T. Train/Scala Books, 1994.

Toperoff, Sholmo Pesach. *The Animal World in Jewish Thought.* Northdale, NJ: Jason Aronson, 1995.

Topsell, Edward and Thomas Muffet. *The History of Four-Footed Beasts and Serpents and Insects* (3 vols.*).* New York: Da Capo, 1967 (facsimile of 1658 edition).

Torga, Miguel. "Vincente the Raven." *Farrusco the Blackbird and Other Stories from the Portuguese.* Translated by Denis Brass. London: George Allen & Unwin, 1950/1941, 83-88.

Tyler, Royall, editor and translator. *Japanese Tales.* New York: Pantheon, 1987.

Veckenstedt, Edmund, editor *Mythen, Sagen und Legenden der Zamiaten* (2 vols.). Heidelberg: Carl Winter's Universitätsbuchhandlung, 1883.

Vest, Jay Hansford C. "From Bobtail to Brer Rabbit: Native American Influences on Uncle Remus." *American Indian Quarterly* 24, no. 1 (2000): 19-43.

Virgil. *The Aeneid.* Translated by Robert Fitzgerald. New York: Random House, 1983.

Virgil. *The Singing Farmer: A Translation of Virgil's "Georgics."* Translated by L. A. S. Jermyn. Oxford: Basil Blackwell, 1947.

Voragine, Jacobus de. *The Golden Legend: Readings on the Saints* (2 vols.). Translated by William Granger Ryan. Princeton, NJ: Princeton University Press, 1995.

Waddell, Helen, editor and translator. *Beasts and Saints.* Grand Rapids, MI: William B. Eerdmans, 1995.

Wade, Nicholas. "The Fly People Make History on the Frontiers of Genetics." *The New York Times.* April 11, 2000, F1-4.

Wang, Chi-Chen. *Traditional Chinese Tales.* New York: Columbia University Press, 1944.

Warner, Marina. *From the Beast to the Blonde: On Fairy Tales and Their Tellers.* New York: Farrar, Straus & Giroux, 1994.

Watson, Henry, translator. *Valentine and Orson.* Edited by Arthur Dickson. New York: Kraus Reprint, 1971.

Webb, Mary. *Precious Bane.* New York: The Modern Library, ca. 1960.

Weinberger, Eliot. "Paper Tiger." Works on Paper 1980-1986. New York: New Directions, 1986, 44-57.

Weinstein, Krystyna, *The Owl in Art, Myth, and Legend.* New York: Crescent Books, 1985.

Wentz, W. Y. Evans. *The Fairy Faith in Celtic Countries.* London: Colin Smythe, 1977.

Werber, Bernard. *Empire of the Ants.* Translated by Margaret Rocques. New York: Bantam, 1999.

Werner, Edward T. C. *Ancient Tales and Folklore of China.* London: Bracken Books, 1986.

White, David Gordon. *Myths of the Dog Man. Chicago, U. of Chicago Press,* 1991.

White, E. B. *Charlotte's Web.* London: Hamish Hamilton Children's Books, 1952.

White, Gilbert. *The Natural History of Selborne.* New York: Frederick Warne, ca. 1895.

White, T. H., translator. *The Book of Beasts: Being a Translation from a Latin Bestiary of the Twelfth Century.* New York: Dover, 1984.

Williams, Ronald J. "The Literary History of a Mesopotamian Fable." *The Phoenix,* vol. 10, #2 (1956), 70-77.

Williams, Terry Tempest. "Undressing the Bear." In *On Nature's Terms.* Edited by Thomas J. Lyon and Peter Stine. Austin: Texas A & M University Press, 1992, 104-107.

Wolkstein, Diane and Samuel Noah Kramer. *Inanna: Queen of Heaven and Earth.* New York: Harper & Row, 1982.

Wood, Rev. J. G. *Man and Beast: Here and Hereafter.* New York: Harper & Brothers, 1875.

Woodbridge, Frederick, James Eugene. *The Son of Apollo: Themes of Plato.* New York: Houghton Mifflin, 1929.

Woolf, Virginia. "The Death of a Moth." *The Death of a Moth and other Essays.* New York: Harcourt, Brace, 1942, 3-6.

Worsted, Donald. *The Wealth of Nature: Environmental History and the Ecological Imagination.* Second edition. New York: Cambridge University Press, 1995.

Wu Ch'eng-en. *Journey to the West* (4 vols.). Translated by Anthony C. Yu. Chicago: University of Chicago Press, 1982.

Yeats, W. B. "Tom O'Roughley." *The Poems of W. B. Yeats.* New York: Macmillan, 1983, 141.

Yoon, Carol Kaesuk. *Naming Nature: The Clash between Instinct and Science.* New York: W. W. Norton, 2009.

Zaehner, R. C. *The Teachings of the Magi: A Compendium of Zoroastrian Beliefs.* New York: Oxford University Press, 1976.

Zimmer, Carl. "Pigeons Get a New Look: In the Bird's Genome, Clues to How Evolution Works." *The New York Times,* February 5 2013, D1, D6.

Zinsser, Hans. *Rats, Lice and History.* New York: Macmillan, 1963.

Ziolkowski, Jan M. *Talking Animals: Medieval Latin Beast Poetry, 750-1150.* Philadelphia: U. of Pennsylvania Press, 1993.

Index of Animals